ADVANCE PRAISE FOR

Critical Pedagogy, Ecoliteracy, & Planetary Crisis

"Richard Kahn has written a dazzling book with the urgency befitting his focus: the ecological crisis that is already upon us, the looming environmental catastrophe worldwide, and the breathtaking arrogance with which powerful economic forces and their political hirelings miseducate, mislead, and misdirect any honest accounting of the mess we're in, or the broad outlines of what is to be done. Kahn aims to shock us awake, to shake us from our deep, deep and sometimes willful sleep of denial, but that is just his opening salvo. His more ambitious project is to contribute to the creation of a mighty and unstoppable social movement geared toward grounded activism on behalf of a humane, balanced, and livable future. Kahn's ethical vision as well as his clear, compelling representation of ecopedagogy and ecoliteracy will change the way we look at education and struggle in and for democracy. This book is essential reading."
—*William Ayers, Distinguished Professor of Education at the University of Illinois at Chicago, and Author of* To Teach *and* Teaching toward Freedom

"This book deals with one of the most important contemporary educational movements: ecopedagogy. In times of crisis convergences—such as the one we are living in, with global warming and profound climatic changes—this book brings an invaluable and accurate contribution, not only to educational theory, but also to the tradition of emancipatory pedagogical practice."
—*Moacir Gadotti, Director, Paulo Freire Institute, São Paulo, Brazil*

"Richard Kahn's *Critical Pedagogy, Ecoliteracy, and Planetary Crisis* is a groundbreaking work that moves the field of critical pedagogy into a visionary mode. Not to address the ecological crisis front and center in these crucial moments of the twenty-first century would be derelict. This work creates a wonderful opening linking critical pedagogy to the emergent scholarship related to ecological literacy."
—*Edmund O'Sullivan, Professor Emeritus, Ontario Institute for Studies in Education*

"Finally, a voice in education that blends critical theory with an ecological ethic inclusive of animal others. Richard Kahn breaks new ground with his ecopedagogy. His book will challenge critical educators to wake up and respond to the times we live in."
—*David Greenwood, Associate Professor, College of Education, Washington State University*

"Here we have education with enlightenment, humanity without hypocrisy, and pedagogy with a punch. Moving from critical pedagogy to ecopedagogy, Kahn transcends the entrenched prejudices and profound limitations of humanism, however radical, for a new educational, ethical, and political paradigm centered on earthlings. He updates pedagogy for the twenty-first century, making it relevant to the social and ecological crises of this profound and unprecedented do-or-die era. This is a supremely important book, the first volley of many to come from one of the most gifted and brilliant thinkers writing today."
      —*Steven Best, Associate Professor of Humanities and Philosophy, University of Texas, El Paso*

"Richard Kahn's continuing scholarship and support for the ecopedagogy movement is an important effort to ensure that environmental education becomes an integral part of the school curriculum. Unfortunately, the Republican and Democratic 2009 national platforms did not mention environmental education and instead emphasized human capital education and its importance for U.S. economic competition in global markets. Human capital approaches to education are the problem because they contribute to the public blindness to the environmental destruction caused by continuing promotion of industrial consumerism. Richard Kahn is fighting the good fight and his ideas and scholarship will help to keep alive efforts to advance the ecopedagogy movement."
      —*Joel Spring, Graduate Center and Queens College, City University of New York*

"Richard Kahn's book hopefully represents the beginning of the end of a long silence on the ecocrisis by the tradition known as critical pedagogy. Importantly, Kahn recognizes that ecology cannot just be tacked on to the list of oppressions that critical pedagogy has concerned itself with, but that it requires that critical pedagogy make itself an object of critique and reconstruction in order to align with the politics of sustainability."
      —*C. A. Bowers, Noted writer and international speaker on educational reforms that address the cultural roots of the ecological crisis*

"Richard Kahn contributes a compelling new voice to debates about how to reimagine a program for North American environmental education. Such a program must redress the marginalization of environmental issues in formal schooling, and do so by bringing a critical, democratic perspective to bear on economic, political, and sociocultural inequality. The current crises in Western modernity make this a rare and opportune moment for students and researchers to explore Kahn's analyses and proposals."
      —*Sandra Harding, Professor, Graduate School of Education and Information Studies, UCLA*

Critical Pedagogy, Ecoliteracy, & Planetary Crisis

Studies in the Postmodern Theory of Education

Joe L. Kincheloe and Shirley R. Steinberg
General Editors

Vol. 359

PETER LANG
New York • Washington, D.C./Baltimore • Bern
Frankfurt am Main • Berlin • Brussels • Vienna • Oxford

RICHARD KAHN

Critical Pedagogy, Ecoliteracy, & Planetary Crisis

THE ECOPEDAGOGY MOVEMENT

PETER LANG
New York • Washington, D.C./Baltimore • Bern
Frankfurt am Main • Berlin • Brussels • Vienna • Oxford

Library of Congress Cataloging-in-Publication Data
Kahn, Richard V.
Critical pedagogy, ecoliteracy, and planetary crisis:
the ecopedagogy movement / Richard Kahn.
p. cm. – (Counterpoints: studies in the postmodern theory of education; v. 359)
Includes bibliographical references and index.
1. Critical pedagogy. 2. Critical thinking. 3. Ecology–Philosophy.
I. Title.
LC196.K344   370.11'5–dc22   2009044525
ISBN 978-1-4331-0545-6
ISSN 1058-1634

Bibliographic information published by **Die Deutsche Nationalbibliothek**.
Die Deutsche Nationalbibliothek lists this publication in the "Deutsche
Nationalbibliografie"; detailed bibliographic data is available
on the Internet at http://dnb.d-nb.de/.

FSC
Mixed Sources
Product group from well-managed
forests, controlled sources and
recycled wood or fiber

Cert no. SCS-COC-002464
www.fsc.org
©1996 Forest Stewardship Council

The paper in this book meets the guidelines for permanence and durability
of the Committee on Production Guidelines for Book Longevity
of the Council of Library Resources.

© 2010 Peter Lang Publishing, Inc., New York
29 Broadway, 18th floor, New York, NY 10006
www.peterlang.com

All rights reserved.
Reprint or reproduction, even partially, in all forms such as microfilm,
xerography, microfiche, microcard, and offset strictly prohibited.

Printed in the United States of America

To my wife, Debbie, who constantly prodded and poked for this book to be completed, and who was always the sustaining force behind its ultimate publication…my love and deepest gratitude for all you do in the world as an educator, as well as for what you must endure, with grace and dignity, as my partner through life.

To my children, Isaiah and Zoë, my cats, Fritz and Kayla, and hounds, Brando and Hanna, as well as to all the beings I have communed with and known intimately and inwardly…my teachers every one.

To all my family and friends of forty years, for all of your constant support and faith, as well as forgiveness of my many flaws…my most humble thanks.

And finally, to those who resist, who are dissatisfied and disobedient, who would rail against the dying of the light and who will not go gently, though they fear not the darkness of this world…this work is especially for you.

Contents

Preface, by Antonia Darder ix

Acknowledgments xix

Ecopedagogy: An Introduction 1

Chapter One
  Cosmological Transformation as Ecopedagogy:
  A Critique of *Paideia* and *Humanitas* 35

Chapter Two
  Technological Transformation as Ecopedagogy:
  Reconstructing Technoliteracy 61

Chapter Three
  The Technopolitics of Paulo Freire and Ivan Illich:
  For a Collaborative Ecopedagogy 81

Chapter Four
  Organizational Transformation as Ecopedagogy:
  Traditional Ecological Knowledge as Real and New Science 103

Chapter Five
  A Marcusian Ecopedagogy 125

Epilogue
  A Concluding Parable: Judi Bari as Ecopedagogue 145

Afterword, by Douglas Kellner
  Mediating Critical Pedagogy and Critical Theory:
  Richard Kahn's Ecopedagogy 151

Bibliography 155

Index 181

Preface

The Great Mother Wails

The Earth extends her arms to us;
revealing through her nature the
changing condition of our existence.

She bends and twists,
deflecting the swords of
our foolishness;
our arrogance;
our gluttony;
our deceit.

Unbridled by red alerts or amber warnings,
Her ire gives rise to monsoon winds,
jarring us from the stupor of
our academic impunity;
our disjointed convolutions,
our empty promises;
our black and white dreams.

Filled with unruly discontent,
we yearn to dominate her mysteries;
reducing her to microscopic dust,
we spit upon her sacredness,
tempting the fury of her seas.

We spill our unholy wars
upon her belly's tender flesh,
blazing dislocated corpses,
ignite her agony and grief.

Still, in love with her creations,
she warns of our complacency

> to cataclysmic devastation,
> rooted in the alienation of
> our disconnection
> our rejection,
> our oppression,
> our scorn.
>
> And still, we spin ungodly
> tantrums of injustice
> against her love,
> against ourselves,
> against one another.
>
> When will we remove blindfolds from our eyes?
> When will we stretch our arms—to her?
> When will the cruelty of our
> hatred cease; teaching us to
> abandon the impositions of
> patriarchy and greed?
>
> Oh! that we might together renew
> our communion with the earth.
> She, the cradle of humanity.
> She, the nourishment of our seeds.
> She, the beauty of our singing.
> She, the wailing that precedes.
>
> —Darder (2008)

It is fitting to begin my words about Richard Kahn's *Critical Pedagogy, Ecoliteracy, and Planetary Crisis: The Ecopedagogy Movement* with a poem. The direct and succinct message of *The Great Mother Wails* cuts through our theorizing and opens us up to the very heart of the book's message—to ignite a fire that speaks to the ecological crisis at hand; a crisis orchestrated by the inhumane greed and economic brutality of the wealthy. Nevertheless, as is clearly apparent, none of us is absolved from complicity with the devastating destruction of the earth. As members of the global community, we are all implicated in this destruction by the very manner in which we define ourselves, each other, and all living beings with whom we reside on the earth.

Everywhere we look there are glaring signs of political systems and social structures that propel us toward unsustainability and extinction. In this historical moment, the planet faces some of the most horrendous forms of "man-made" devastation ever known to humankind. Cataclysmic "natural disasters" in the last decade have sung the environmental hymns of planetary imbalance and reckless environmental disregard. A striking feature of this ecological crisis, both locally and globally, is the overwhelming concentration of wealth held by the ruling elite and their agents of capital. This environmental malaise is characterized by the staggering loss of livelihood among working people everywhere; gross inequalities in educational opportunities; an absence of health care for millions; an unprecedented number of people living behind bars; and trillions spent on fabricated wars fundamentally tied to the control and domination of the planet's resources.

The Western ethos of mastery and supremacy over nature has accompanied, to our detriment, the unrelenting expansion of capitalism and its unparalleled domination over all aspects of human life. This hegemonic worldview has been unmercifully imparted through a host of public policies and practices that conveniently gloss over gross inequalities as commonsensical necessities for democracy to bloom. As a consequence, the liberal democratic rhetoric of "we are all created equal" hardly begins to touch the international pervasiveness of racism, patriarchy, technocracy, and economic piracy by the West, all which have fostered the erosion of civil rights and the unprecedented ecological exploitation of societies, creating conditions that now threaten our peril, if we do not reverse directions.

Cataclysmic disasters, such as Hurricane Katrina, are unfortunate testimonies to the danger of ignoring the warnings of the natural world, especially when coupled with egregious governmental neglect of impoverished people. Equally disturbing, is the manner in which ecological crisis is vulgarly exploited by unscrupulous and ruthless capitalists who see no problem with turning a profit off the backs of ailing and mourning oppressed populations of every species—whether they be victims of weather disasters, catastrophic illnesses, industrial pollution, or inhumane practices of incarceration. Ultimately, these constitute ecological calamities that speak to the inhumanity and tyranny of material profiteering, at the expense of precious life.

The arrogance and exploitation of neoliberal values of consumption dishonor the contemporary suffering of poor and marginalized populations around the globe. Neoliberalism denies or simply mocks ("Drill baby drill!") the interrelationship and delicate balance that exists between all living beings, including the body earth. In its stead, values of individualism, competition, privatization, and the "free market" systematically debase the ancient ecological knowledge of indigenous populations, who have, implicitly or explicitly, rejected the fabricated ethos of "progress and democracy" propagated by the West. In its consuming frenzy to gobble up the natural resources of the planet for its own hyperbolic quest for material domination, the exploitative nature of capitalism and its burgeoning technocracy has dangerously deepened the structures of social exclusion, through the destruction of the very biodiversity that has been key to our global survival for millennia.

Kahn insists that this devastation of all species and the planet must be fully recognized and soberly critiqued. But he does not stop there. Alongside, he rightly argues for political principles of engagement for the construction of a critical ecopedagogy and ecoliteracy that is founded on economic redistribution, cultural and linguistic democracy, indigenous sovereignty, universal human rights, and a fundamental respect for all life. As such, Kahn seeks to bring us all back to a formidable relationship with the earth, one that is unquestionably rooted in an integral order of knowledge, imbued with physical, emotional, intellectual, and spiritual wisdom. Within the context of such an ecologically grounded epistemology, Kahn uncompromisingly argues that our organic relationship with the earth is also intimately tied to our struggles for cultural self-determination, environmental sustainability, social *and* material justice, and global peace.

Through a carefully framed analysis of past disasters and current ecological crisis, Kahn issues an urgent call for a critical ecopedagogy that makes central explicit articulations of the ways in which societies construct ideological, political, and cultural systems, based on social structures and practices that can serve to promote ecological sustainability and biodiversity or, conversely, lead us down a disastrous path of unsustainability and extinction. In making his case, Kahn provides a grounded examination of the manner in which consuming capitalism manifests its repressive force throughout the globe, disrupting the very ecological order of knowledge essential to the planet's sustainability. He offers an understanding of critical ecopedagogy and ecoliteracy that inherently critiques the history of Western civilization and the anthropomorphic assumptions that sustain patriarchy and the subjugation of all subordinated living beings—assumptions that continue to inform traditional education discourses around the world. Kahn incisively demonstrates how a theory of multiple technoliteracies can be used to effectively critique the ecological corruption and destruction behind mainstream uses of technology and the media in the interest of the neoliberal marketplace. As such, his work points to the manner in which the sustainability rhetoric of mainstream environmentalism actually camouflages wretched neoliberal policies and practices that left unchecked hasten the annihilation of the globe's ecosystem.

True to its promise, the book cautions that any anti-hegemonic resistance movement that claims social justice, universal human rights, or global peace must contend forthrightly with the deteriorating ecological crisis at hand, as well as consider possible strategies and relationships that rupture the status quo and transform environmental conditions that threaten disaster. A failure to integrate ecological sustainability at the core of our political and pedagogical struggles for liberation, Kahn argues, is to blindly and misguidedly adhere to an anthropocentric worldview in which emancipatory dreams are deemed solely about human interests, without attention either to the health of the planet or to the well-being of all species with whom we walk the earth.

Important to the contributions of this volume is the manner in which Kahn retains the criticality of the revolutionary project in his efforts to dialectically engage the theories of Paulo Freire and Ivan Illich, in ways that significantly pushes Freire's work toward a more ecologically centered understanding of human liberation and that demonstrates Illich's continued relevance on these matters. Key to his argument is the recognition of planetary sustainability as a vital and necessary critical pedagogical concern. In a thoughtful and effective manner (which has been long coming), Kahn counters spurious criticisms railed against the integrity of critical pedagogy and its proponents. Instead, he highlights both the radical underpinnings of critical theoretical principles and the historicity of its evolution—acknowledging both its significant contributions to the field, as well as its shortcomings in past articulations. Rather than simply echo denouncements of "beyond critical pedagogy," Kahn intricately weaves possibilities drawn from Freire and Illich, neither essentializing the work of these theorists nor ignoring the problematic instances of their formulations. This discussion brings a mature and refreshing sense of both political grace and sober critique, which supports the passion of our pedagogical traditions, while simultaneously chastising our slowness in taking up the mantle of ecological responsibility.

Through the reformulation of Herbert Marcuse's contributions to critical theories of society, Kahn gives voice to a North American ecopedagogy that thoughtfully seizes the power of radical environmental activists, while simultaneously opposing and calling for the remaking of capitalist ecological practices, as a key component to any critical pedagogical project. By so doing, critical pedagogy is forcefully challenged to step up to the demands and needs of a world in ecological crisis, in the hopes of transforming itself into a counter-hegemonic resistance movement imbued with ecological consciousness, respect for beauty in all life, and a serious commitment to preserving the multifarious nature of our humanity. In the process, Kahn propels us beyond the debilitating theoretical posturing of the left in ways that liberate our political sensibilities and guide us toward alternative pedagogies of knowledge construction and new technopolitics of education necessary for our future sustainability.

Similar to revolutionary ecologists before him, Kahn urges for a critical shift in our worldview from one that is dominated by the instrumentalization of ethnocentrism, xenophobia, militarism, and the fetishizing of all living functions, to one that acknowledges unapologetically and wholeheartedly the deep intimacy and organic connection at work in all forms of existence. In the spirit of Vandana Shiva's "earth democracy," Kahn also argues for a ecopedagogy that demands we "remove our blinders, imagine and create other possibilities," reminding us that "Liberation in our genocidal times, is, first and foremost, the freedom to stay alive."[1]

True to this dictum, Kahn unambiguously demands that the survival of the planet (and ourselves!) underscore our political and pedagogical decisions, despite the fact that seldom have questions of ecological concern been made central to the everyday lives of teachers and students or to the larger context of movement work, save for the liberal agenda of the Sierra Club or the well-meaning discourse on population control for poor and racialized women, espoused by people of all ideological stripes. Perhaps, it is this "missing link" in the curriculum of both public schools and political movements that is most responsible for the historically uncritical and listless response to the global suffering of human beings subjected to imperial regimes of genocide, slavery, and colonialism. In truth, a deeper analysis exposes sharply a legacy that persists today in the shrouded values and attitudes of educators from the dominant class and culture who expect that all oppressed populations and living species should acquiesce to the dominion and hegemonic rule of the wealthy elite.

It is precisely such a worldview of domination that perpetuates the extinction of whole species, as it does the cultural and linguistic destruction of peoples and nations outside of a "first-world" classification. As a consequence, our biodiversity is slipping away, despite scientific findings that clearly warn of the loss of hardiness and vitality to human life, as a direct consequence of the homogenization of our differences. It is equally ironic to note here how repression of the body itself is manifested within the capitalist fervor to commodify or colonize all forms of vital existence. Schools, unfortunately, are one of the most complicit institutions in the exercise of such ecological repression, generally carried out through the immobilization of the body and the subordination of our emotional nature, our sexual energies, and spiritual capacities.

In response, Kahn eloquently argues for a critical ecopedagogy and ecoliteracy that supports teachers in engaging substantively students' integral natures, in an effort to forge an emancipatory learning environment where all can thrive amid everyday concerns. As such, he makes clear that, although important, it is not enough to rely solely on abstract cognitive processes, where only the analysis of words and texts are privileged in the construction of knowledge. Such an educational process of estrangement functions to alienate and isolate students from the natural world around them, from themselves, and one another. This, unwittingly, serves to reinforce an anthropocentric reading of the world, which denies and disregards the wisdom and knowledge outside Western formulations. In contrast, an ecopedagogy that sustains life and creativity is firmly grounded in a material and social understanding of our interconnected organic existence, as a starting place for classroom practice and political strategies for reinventing the world.

Also significant to Kahn's notion of ecopedagogy is an engagement with the emancipatory insights and cultural knowledge of indigenous populations, given that the majority of the social and political problems facing us today are fundamentally rooted in mainstream social relations and material conditions that fuel authoritarianism, fragmentation, alienation, violence, and greed. Such anti-ecological dynamics are predicated on an ahistorical and uncritical view of life that enables the powerful to abdicate their collective responsibility to democratic ideals, while superimposing a technocratic and instrumental rationality that commodifies and objectifies all existence. Such a practice of education serves to warp or marginalize diverse indigenous knowledge and practices, by privileging repetitive and unimaginative curricula and fetishized methods. Anchored upon such a perspective of schooling, classroom curriculum socializes students into full-blown identities as entitled consuming masters and exploiters of the earth, rather than collective caretakers of the planet.

In contrast, Kahn explores the inherent possibilities at work within indigenous knowledge and traditions, in ways that enhance our capacity to not only critique conditions of ecological crisis, but to consider ways in which non-Western societies and peoples have enacted ecologically sustaining practices within the everyday lives of their communities. He turns the false dominion of the West on its head, offering alternative ways of being that hold possibilities for the reconstruction of institutional culture, the transformation of how we view technology and science, and thus the reformulation of public policy. As critical educators and revolutionary activists across communities of difference, we are encouraged to turn to the wisdom of our own historical survival, in serious and sustained ways, in order to work toward the abandonment of colonizing values and practices that for centuries have denigrated our cultural ways and attempted to disable our life-sustaining capacities.

Moreover, to contend effectively with issues of racism, sexism, homophobia, disablism, and other forms of inequalities, a life-affirming ecological praxis is paramount. That is, one that encompasses a refusal to adhere to political, economic, and philosophical disconnections, which falsely separate humankind from those ecological dynamics that shape local, global, regional, rural, and urban landscapes. Instead, static views of humanity and the planet, which inadvertently serve the commodifying interests of capital and its penchant to divide and conquer, are challenged and dismantled through an integral political solidarity of heart, mind, body, and spirit. Accordingly, a critical ecopedagogy must then encompass those philosophical principles that are at home with ambiguity, dissonance, difference, and heterogeneity, as an ever-present phenomenon. Such an ethos supports a world where cross-species concerns are both commonplace and valued for their creative potential in the making of a truly democratic, just, and peaceful world.

At the heart of Kahn's project is the intention to move us beyond a capitalist orthodoxy of consumerism, careerism, and corporate profiteering. As educators, we are invited to commit ourselves to a critical ecopedagogy that courageously embraces a new paradigm for the living out of a transformative ecological praxis—one that is shaped by the power of human emotions, the cultural rituals of diverse ways of being, a deep respect for universal rights, and the integration of planetary consciousness. More importantly, he points us toward re-envisioning ourselves as activists, committed to ending oppression in all its manifestations, through embracing with revolutionary love and grace the significance and necessity of all life forms.

The late Murray Bookchin, in *The Ecology of Freedom*, proclaimed that "Humanity has passed through a long history of one-sidedness and of a social condition that has always contained the potential of destruction, despite its creative achievements in technology. The great project of our time must be to open the other eye: to see all-sidedly and wholly, to heal and transcend the cleavage between humanity and nature that came with early wisdom."[2] True to these words, Kahn urges us "to open our other eye" and be mindful of the delicate balance of the earth and our collective accountability to future generations. Written with analytical prowess, uncompromising courage, and political fortitude, *Critical Pedagogy, Ecoliteracy, and Planetary Crisis: The Ecopedagogy Movement* draws upon the passion of revolutionary visions and ancient indigenous sensibilities to awaken us to our responsibility and unequivocal commitment to the sustainability of all life. Through the perseverance of his own political and pedagogical reflections, Richard Kahn invites us to discover the beauty of a steadfast ecology of life—one that might help to release us from the bondage of our inhumanities.

> *When we've totally surrendered to that beauty,*
> *We'll become a mighty kindness.*
> —Rumi

Professor Antonia Darder
Distinguished Professor of Education
University of Illinois, Urbana Champaign

NOTES

1. See Vandana Shiva (2005), *Earth Democracy Justice, Sustainability, and Peace*, Boston: South End Press: p. 185.

2. See Murray Bookchin's *The Ecology of Freedom* (2005), Oakland, CA: AK Press: p. 152.

Acknowledgments

A very special thanks is owed to Douglas Kellner, my doctoral mentor, frequent co-author, and supplier of this book's Afterword. It is common in our educational circles to speak of transformation. Truly, our relationship has transformed my life. You gave me the break I needed and it was under your tutelage that I have become the scholar that I am today. Always kind, gracious, and wise beyond belief, I appreciate immensely all that you have done for me, as I strive now to model the same for my own doctoral students.

Also, let me express profound thanks to Steve Best and Peter McLaren. Your friendship and influence have meant a great deal to me over the years, and I have learned much from you both. Perhaps, more than anything else, I have learned that my work should strive to be courageously relevant to the historical situation at hand, and that sometimes a true comrade is as powerful and necessary as a new idea. No small teaching indeed.

Finally, a big hug to Antonia Darder for her wonderful Preface for this book, delivered under challenging conditions. I will always remember meeting for the first time at the Paulo Freire Forum at UCLA, under what seemed like the kind of magical atmosphere that only artists, or people in tune with spiritual realms, or lovers, are aware of and can admit. You amaze and inspire. To be on your side in this struggle makes me proud.

Let me gratefully acknowledge all the fantastic scholars who have lent support to the ecopedagogy project over the years, including (but not limited to): Shirley Steinberg, Joe Kincheloe, Ken Saltman, Chet Bowers, Bill Ayers, Carlos Torres, Joel Spring, Anthony Nocella, Greta Gaard, Pauline Sameshima, Henry Giroux, David Greenwood, Madhu Suri Prakash, Moacir Gadotti, Sandra Harding, Jenny Sandlin, Tyson Lewis, Clayton Pierce, Dolores Calderon, Nathalia Jaramillo, Dave Hill, Gene Provenzo, Connie Russell, Helena Pedersen, Simon Boxley, Edmund O'Sullivan, Therese Quinn, Rebecca Martusewicz, Peter Mayo, Ilan Gur-Ze'ev, Jason Lukasic, Rhonda Hammer, Levana Saxon, Donna Houston, Greg Martin, Peter Buckland, all those who have been involved in the *Green Theory & Praxis* journal, my phenomenal colleagues in the Department of Educational Foundations & Research at UND, Isham Christie and UND's Students for a Democratic Society, the members of my 2009 Foundations of Ecoliteracy class, and the many, many others who undoubtedly deserve mention here save that space prevents me from more properly articulating my sincere appreciation for your assistance, ideas, and influence on my work.

I also acknowledge with admiration all of the many different stripes of activists working for another world on a daily basis, especially those who exhibit the courage to take direct action in building the foundations of a better reality. Jerry Vlasak, in particular, has always been a good friend and great speaker in my college classes. It has been my privilege also to know Kevin Jonas, and my thoughts go out to him while he serves time in a Minnesota prison for his leadership in the highly successful SHAC-USA campaign. Vivisectors all should rot in the hell they are daily creating.

Special thanks are owed to David Ulansey for teaching me of the Sixth Extinction during my time at the California Institute of Integral Studies.

Again, my gratitude goes to Shirley (and Joe) for accepting this book for the Counterpoints series; and thanks to Sophie Appel and all those at Peter Lang who assisted me in the production and publication of this manuscript.

Lastly, I should note that while this book has been substantially rewritten in both form and content, it is based on my dissertation work completed at the University of California, Los Angeles, in 2007, under the title, *The Ecopedagogy Movement: From Global Ecological Crisis to Cosmological, Technological, and Organizational Transformation of Education*.

Elements of that dissertation appeared in the following publications:

Kahn, R. Toward a Critique of Paideia and Humanitas: (Mis)Education and the Global Ecological Crisis. In I. Gur-Ze'ev & K. Roth (Eds.), *Education in the Era of Globalization*. New York: Springer, 2007. All rights reserved.

Kahn, R. & Kellner, D. Reconstructing Technoliteracy: A Multiple Literacies Approach. *E-Learning* 2(3). Symposium Journals, 2005. All rights reserved.

Kahn, R. & Kellner, D. Paulo Freire and Ivan Illich: Technology, Politics and the Reconstruction of Education. In C. Torres & P. Nogeura (Eds.), *Social Justice Education for Teachers: Paulo Freire and the Possible Dream*. The Netherlands: Sense Publishers, 2008. All rights reserved.

Kahn, R. The Educative Potential of Ecological Militancy in an Age of Big Oil: Towards a Marcusian Ecopedagogy. *Policy Futures in Education* 4(1). Symposium Journals, 2006. All rights reserved.

Ecopedagogy: An Introduction

> Even the most casual reading of the earth's vital signs immediately reveals a planet under stress. In almost all the natural domains, the earth is under stress—it is a planet that is in need of intensive care. Can the United States and the American people, pioneer sustainable patterns of consumption and lifestyle, (and) can you educate for that? This is a challenge that we would like to put out to you.
> —Noel J. Brown, United Nations Environment Programme (in Ince, 1995, p. 123)

> Our destiny exercises its influence over us even when, as yet, we have not yet learned its nature: it is our future that lays down the law to our today.
> —Friedrich Nietzsche (1908)

Introducing the Problem

In 1970, the first Earth Day event helped to mark the global arrival of the environmental movement and it is often hailed as a pedagogical and political milestone toward the production of a more ecologically sound society. By contrast, it is not uncommon today to hear students, environmentalists, and other informed citizens criticize Earth Day with declarations like, "Every day should be Earth Day—to give the Earth one day a year of love and respect, while denying it the other 364 doesn't help much at all." While such critique can be symptomatic of a form of paralyzing and reactionary cynicism, it should also be seen as representative of modern environmentalism's compelling achievement as an educational social movement to date. Whereas the critical socioenvironmental visions of theorists such as Aldo Leopold, Rachel Carson, or Murray Bookchin must have sounded like voices crying out in the wilderness in the 1950s or 1960s, in the twenty-first century it is no longer necessary for a great many people to argue even about the ecological burdens produced by global society. However, if recent decades have seen the rise of a powerful popular demand for planetary sustainability, this must be placed in the alarming context of the more rapid expansion of unsustainable economic practices throughout the world since the end of World War II—

the modern development strategies commonly denoted by the discourse of "globalization."

In 2005, the UN-funded Millennium Ecosystem Assessment (MEA) released the most encompassing study to date about the state of the planet's ecology. To summarize, it found that during the last fifty years, humanity has altered (and mainly degraded) the earth's ecosystems "more rapidly and extensively than in any comparable time of human history" (MEA, 2005, p. 2). This was done largely on behalf of an exponential demand for primary natural resources that coincides with the social and economic changes wrought by corporate and other transnational capitalist interests (Kovel, 2007). For instance, between 1960 and 2000, the world's population doubled and the global economy increased by more than sixfold. At the same time, the mining of and dependence upon large-scale industrial energy resources like oil, coal, and natural gas followed and exceeded the trends set by the population curve despite many years of warnings about the consequences inherent in their overuse and extraction. This, of course, has led to a corresponding increase in the carbon emissions known to be responsible for global warming (Gore, 2006).[1]

Additionally, more land (e.g., forests, wetlands, prairies, savannahs) has been converted for agricultural uses over the last half-century than had taken place during the 150 years prior combined (MEA, 2005, p. 2). The majority of the world's dominant farming practices (e.g., agribusiness monocropping; slash-and-burn technique) developed during this period has debased soil quality and furthered global desertification. However, the so-called "green revolution" has been sold as a success because short-term food production via these methods increased by a factor of nearly three. Other land usage statistics from this time frame show that water use doubled (nearly 70 percent of used water goes to agriculture), half of all wetlands were developed, timber pulping and paper production tripled while 50 percent of the forests disappeared, and the damming of flowing waterways doubled hydropower (p. 5). Moreover, unsustainable fishing practices contributed to grave losses of global mangroves during the second half of the twentieth century, reducing them by approximately 35 percent. Coral reef biomes—our underwater tropical rain forests—have likewise tolled worldwide extinction and damage rates of 20 percent each respectively since 1960 (p. 5).

This has led (and will continue to lead) to unthinkable levels of marine species extinction. The rise of commercial fishing is now known to have eradicated some 90 percent of the ocean's largest fish varieties. Forty-mile-long drift nets are routinely used to trawl the ocean bottoms, causing incalculable damage to the ocean ecosystem. Giant biomass nets, with mesh so fine that not even baby fish can escape them, have become the industry standard in commercial fishing and, as a result, there is expected to be no extant commercial fishery left active in the world by 2048 (Worm, et al., 2006). Further, such nets are commonly drowning and killing about 1,000 whales, dolphins, and porpoises daily, some of the very species already near extinction from centuries of commercial hunting (Verrengia, 2003), and there has even been a startling move toward the reintroduction of commercial whaling by the International Whaling Commission due to pressure from countries such as Norway, Iceland, and Japan.

The effects of corporate globalization have been equally profound on other species, as we have experienced 1,000 times the historical rate of normal background extinction, with upwards of 30 percent of all mammals, birds, and amphibians currently threatened with permanent disappearance (MEA, 2005, p. 4). In other words, over the span of just a few decades we are involved in a mass die-off of nonhuman animals such as we have not witnessed for 65 million years, and worse yet, predictions for the future expect these rates of extinction to increase tenfold (p. 5). Moreover, these figures only document the indirect destruction of land animals and so fail to account for the ways in which capitalism has transformed family farms and subsistence-oriented agriculture into vast, unimaginable factory farms and their corresponding slaughterhouses—brutal and ecologically ruinous production lines, in which thousands of animals are murdered for meat harvesting every hour per the business standard (Singer & Mason, 2006).

Almost all of these trends just summarized are escalating and most are accelerating. Even during what amounts to a current economic downturn, transnational markets and neoliberal policies continue to flow and evolve, and the globalization of technocapital (Best & Kellner, 2001) persists in order to fuel yet another vast reconstruction of the information society that has developed under the aegis of American imperialism. Over the last fifty to sixty years, then, a particularly noxious economic paradigm has unfolded like a shock wave across the face of the earth, one that has led to an exponential increase of global capital and startling achievements in science and technology, but which has also had devastating effects upon ecosystems both individually and taken as a whole (Foster, 2002). According to the United Nations Environment Programme's *GEO-3* report, a vision of continued economic growth of this kind is consonant only with planetary extinction:

either great changes are made in our global lifestyle now or irrevocable social and ecological upheavals will grip the world by 2032 (United Nations Environment Programme, 2002).

Ecocrisis and Environmental Education

> Nor do piecemeal steps however well intended, even partially resolve problems that have reached a universal, global and catastrophic character. If anything, partial "solutions" serve merely as cosmetics to conceal the deep seated nature of the ecological crisis. They thereby deflect public attention and theoretical insight from an adequate understanding of the depth and scope of the necessary changes.
> —Murray Bookchin (1982)

For the reasons just outlined, many now routinely speak of an unprecedented global environmental or ecological crisis (or crises) as being underway. However, while the term *crisis* is utilized in a colloquial fashion to connote ideas of uncontrollable mayhem and danger, it should rather be understood as a diagnostic philosophical concept that indicates the need for personal critical deliberation toward the possibility of affecting meaningful change. Etymologically, the concept relates to the ancient Greek verb *krinein*, which means "to decide." Throughout history, the idea of crisis has also possessed a primary medical connotation in which it identifies the potential turning point of diseases in which the infirm will either begin to gain health or become fatally ill. This diagnostic aspect of the term doubtlessly informed its use as a modern political concept beginning during the Age of Enlightenment when revolutionary activity, sociocultural disruptions, and sweeping changes in the economy led to the creation of new theories and intellectual perspectives in the attempt to reveal the symptoms of social pathology and provide prognoses that might ensure a better future. Hence, to be subjected to crisis is to partake of structural threats and potential failures but it is also, contradictorily, to be able to identify threats such that they become the objects of one's own autonomous decision-making power. A crisis should thus be seen as "a moment of decisive intervention…of thorough-going transformation…of rupture" (Hay, 1999, p. 323). It is potentially catastrophic, but not necessarily so—the matter very much hangs in the balance. The idea is captured succinctly by Frijtof Capra, who noted in the opening of his own founding ecological manifesto, *The Turning Point* (1984), that the Chinese ideogram "for 'crisis' - *wei-ji* - is composed of the characters for 'danger' and 'opportunity'" (p. 26).

Just as there is now an ecological crisis of serious proportions, there is also a crisis in environmental education over what must be done about it. Again, over the last half-century, the modern environmental movement has undeniably helped to foster widespread social and cultural transformation. In part, it has developed ideas and practices of environmental preservation and conservation, struggled to understand and reduce the amount of pollution and toxic risks associated with industrialized civilization, produced new modes of counterculture and morality, outlined the need for appropriate technologies, and led to powerful legislative environmental reforms as well as a wide range of alternative institutional initiatives. As a form of nonformal popular education it has stirred many people to become self-aware of the role they play in environmental destruction and to become more socially active in ways that can help to create a more ecological and sustainable world.

In terms of formal educational programs, federal and state legislatures have mandated that environmental education be included as part of the public education system's curricular concerns. Over the last thirty-eight years, the North American Association for Environmental Education—the world's flagship environmental education organization—has grown from being a fledgling professional society to its current state as the coordinator, in over fifty-five countries worldwide, of thousands of environmental organizations toward the certification and legitimation of environmental education as a professional research field. These educational programs have apparently made their case, as a comprehensive set of studies completed in 2005 found that:

- 95% of all American adults support having environmental education programs in schools;

- 85% of all American adults believe that governmental agencies should support environmental education programs; and that

- 80% believe that corporations should train their employees in how to solve environmental problems. (Coyle, 2005)

In many ways, then, the foundation for comprehensive and powerful forms of environmental literacy and ecoliteracy has never been more at hand throughout society.

To reiterate: despite the environmental movement's significant pedagogical accomplishments, there have also been numerous setbacks and a tremendous amount of work remains to be done—perhaps more than ever before (see the still relevant Dowie, 1996). For example, the same studies that revealed Americans' overwhelming support for environmental education programs reported a variety of findings which demonstrate that most Americans continue to have an almost shameful misunderstanding of the most basic environmental ideas. Thus, it was found that an estimated:

- 45 million Americans think the ocean is a fresh source of water;

- 125 million Americans think that aerosol spray cans still contain stratospheric ozone-depleting chlorofluorocarbons (CFCs) despite the fact that they were banned from use in 1978;

- 123 million Americans believe that disposable diapers represent the leading landfill problem when they in fact only represent 1% of all landfill material; and

- 130 million Americans currently believe that hydropower is the country's leading energy source when, as a renewable form of energy, it contributes only 10% of the nations total energy supply. (Coyle, 2005)

Of course, more problematic still for educators is the burgeoning rise in social and ecological disasters that are resulting from the mixture of unsustainable economic exploitation of nature and environmentally unsound cultural practices.[2] Such ecological issues, requiring critical knowledge of the dialectical relationship between mainstream lifestyle and the dominant social structure, require a much more radical and more complex form of ecoliteracy than is presently possessed by the population at large. In this context, while it may be unfair to lay the blame for social and ecological calamity squarely on the environmental movement for its inability to generate effective pedagogy on this matter, it must still be noted that the field of environmental education has been altogether unable to provide either solutions or stop-gaps for the ecological disasters that have continued to mount due to the mushrooming of transnational corporate globalization over the last few decades.

In fact, despite a proliferation of programs since the 1970s, environmental education has tended to become isolated as a marginal academic discipline relative to the curricular whole.[3] The major trend on campuses today is for environmental studies to be lodged within and controlled by natural sciences departments, with little more than tips of the cap to the humanities, and ostensibly no input from scholars of education (see Kahn &

Nocella, Forthcoming). When such studies are housed in colleges of education proper, however, they are rarely integrated across required programs of study in either teacher training, educational leadership, or educational research. Instead, they are generally confined to specialized M.A.-level or other certificate-based environmental education programs.

These degree programs often lack rigorous training in theoretical critique and political analysis, choosing to focus instead on the promotion of outdoor educational experiences that all too often advance outdated, essentialized, and dichotomous views of nature and wilderness.[4] As Steven Best and Anthony Nocella (2006) have argued, such views as these are typical of the first two waves of (predominantly white, male, and middle-class) U.S. environmentalism. These views have proven insufficient and even harmful toward the advancement of richly multiperspectival ecological politics and environmental justice strategies (for instance, see Adamson, et al., 2002), which seek to uncover collective social action across differences of race, class, gender, species, and other social categories. Hence, many outdoor education programs stand in need of radical reconstruction away from an uncritical form of environmental literacy that has remained rooted as the field standard since William Stapp (1969). Stapp is considered the "founder" of the environmental education movement. He first stressed that the goals of environmental education were: knowledge of the natural environment, interdisciplinary exploration, and an inquiry-based, student-centered curricular framework, which could be used for overcoming intractable conflict and ideology in society.[5]

Critiquing Environmental Literacy: The Zoo School

A poster-child example for such environmental literacy[6] is the School of Environmental Studies, known as the "Zoo School," in Apple Valley, Minnesota. Here high school-aged juniors and seniors attend school on the zoo grounds, treating the institution and a nearby park as an experiential learning lab where they conduct independent studies and weave environmental themes into their curricular work and projects. A recent pamphlet funded and promoted by the U.S. Environmental Protection Agency, Office of Environmental Education, *Advancing Education Through Environmental Literacy* (Archie, 2003) lauds the school as one "using the environment to boost academic performance, increase student motivation, and enhance environmental literacy" (p. 8). But the literacy aspects of this education, which accord with the aims put forth by Stapp and those of the North American Association for Environmental Education, lack the strong critical and ethical focus that is presently demanded by our unfolding planetary ecocrisis.[7]

For example, per written accounts, the heads of the Zoo School do not have the students pose problems into the history and nature of zoos—a highly problematical social and environmental institution (Rothfels, 2002)—or become active in the fight against the Apple Valley zoo's own sordid history and policies. As regards the latter project, a worthwhile educational venture would be to have students become involved in banning dolphins as a zoo exhibit (hardly a native species to Minnesota) and to have them returned to either a sanctuary or non-domesticated oceanic habitat. Instead, as of 2006, one could pay $125 to swim with the zoo's dolphins, a practice generally condemned by marine ecologists (Rose, 1996) and environmentalists/animal rightists (Watson, 1995) alike as both inhumane and beyond the bounds of good environmental stewardship.

Further, the Apple Valley zoo's Wells Fargo Family Farm claims to foster environmental literacy experiences for Zoo School students "to explain and…learn about how food gets from farms to tables."[8] Yet students could alternatively work for a critical literacy that seeks to understand how the implosion of corporate marketing and ideology into the zoo structures its educational program. That is, while the Zoo School presently offers relatively idealized experiences of life on a family farm, it could instead aim for literacy into how to organize opposition to such questionable practices as the naturalization of a corporate "family farm," as well as in how to demand answers from responsible parties as to why high-ranking executives of a leading corporate agribusiness like Cargill presently sit on the zoo's board of directors. Additionally, students could learn to read the corporate farm exhibit against the grain in order to politically problematize why the zoo has failed to create educational encounters on the ecological benefits of a vegan diet, when it instead at least tacitly supports as sustainable and conservationist-minded the standard American meat-based diet and the ecologically damaging factory farming that presently supports it.

Failing to provide critical pedagogy, the Zoo School has been promoted within leading environmental education circles as a leader because it is, in the words of the Environmental Education & Training Partnership, "Meeting Standards Naturally" (Archie, 2003). That is, it is motivating students in a new way to go to school and meet or even surpass national curricular and testing standards of a kind consistent with the outcome-orientation of the No

Child Left Behind Act. As with other schools that have adopted environmental education as the central focus of their programs, the Zoo School apparently shines—not because it is producing ecological mindsets and sustainable living practices capable of transforming society in radically necessary ways—but because its students' reading and math scores have improved; and they have performed better in science and social studies; developed the ability to transfer their knowledge from familiar to unfamiliar contexts; learned to "do science" and not just learn about it; and showed a decline in the sort of overall behavior classified as a discipline problem (Glenn, 2000, p. 3). Obviously, regardless of whatever good pedagogy is taking place at the Zoo School, this laudatory praise of its environmental literacy program by environmental educators is little more than the present-day technocratic standards movement in education masquerading as a noteworthy "green" improvement. Put bluntly: this is environmental literacy as a greenwash.[9]

Worse still, though, is that here environmental literacy has not only been co-opted by corporate state forces and morphed into a progressively-styled, touchy-feely method for achieving higher scores on standardized tests like the ACT and SAT, but in an Orwellian turn it has come to stand in actuality for a real illiteracy about the nature of ecological catastrophe, its causes, and possible solutions. As I will argue in this book, our current course for social and environmental disaster (though highly complex and not easily boiled down to a few simple causes or strategies for action) must be traced to the evolution of: an anthropocentric worldview grounded in what the sociologist Patricia Hill Collins (1993) refers to as a matrix of domination (see chapter 1); a global technocapitalist infrastructure that relies upon market-based and functionalist versions of technoliteracy to instantiate and augment its socio-economic and cultural control (see chapters 2 and 3); an unsustainable, reductionistic, and antidemocratic model of institutional science (see chapter 4); and the wrongful marginalization and repression of pro-ecological resistance through the claim that it represents a "terrorist" force that is counter to the morals of a democratic society rooted in tolerance, educational change, and civic debate (see chapter 5). By contrast, the environmental literacy standards now showcased at places like the Zoo School as "Hallmarks of Quality" (Archie, 2003, p. 11) are those that consciously fail to develop the type of radical and partisan subjectivity in students, that might be capable of deconstructing their socially and environmentally deleterious hyper-individualism or their obviously socialized identities that tend toward state-sanctioned norms of competition, hedonism, consumption, marketization, and forms of quasi-fascistic patriotism.

Just as Stapp (1969) theorized environmental literacy as a form of political moderation that could pacify the types of civic upheaval, that occurred during the Civil Rights era, now too during the tendentious political atmosphere that has arisen as the legacy of the George W. Bush presidency, being environmentally literate quite suspiciously means learning how to turn the other cheek and listen to "both sides" of an issue—even when the issue is the unprecedented mass extinction of life taking place on the planet. In a manner that accords more with Fox News than Greenpeace, a leading environmental literacy pamphlet (Archie, 2003) emphasizes that "Teaching and learning about the environment can bring up controversies that must be handled in a fair and balanced manner in the classroom" (p. 11). Later in the document a teacher from Lincoln High School in Wisconsin is highlighted in order to provide expert advice in a similar fashion: "I'd say the most important aspect of teaching about the environment is to look at all aspects involved with an issue or problem. Teach from an unbiased position no matter how strong your ideas are about the topic. Let the kids make decisions for themselves" (p. 12), she implores.

This opinion is mirrored by the Environmental Education Division of the Environmental Protection Agency (a federal office, created by the Bush administration, dedicated to furthering environmental literacy), which on its own website underscores as "Basic Information" that "Environmental education does not advocate a particular viewpoint or course of action. Rather, it is claimed that environmental education teaches individuals how to weigh various sides of an issue through critical thinking and it enhances their own problem-solving and decision-making skills."[10] Yet, this definition was authored by an administration trumping for a wider right-wing movement that attempts to use ideas of "fair and balanced" and "critical thinking" to occlude obvious social and ecological injustices, as well as the advantage it gains in either causing or sustaining them. This same logic defending the universal value of nonpartisan debate has been used for well over a decade by the right to prevent significant action on global warming. Despite overwhelming scientific acceptance of its existence and threat, as well as of its primarily anthropogenic cause, those on the right have routinely trotted out their own pseudo-science on global warming and thereby demanded that more research is necessary to help settle a debate on the issue that only they are interested in continuing to facilitate.

Likewise, within academic circles themselves, powerful conservatives like David Horowitz have the support of many in government who are seeking to target progressive scholars and viewpoints on university and college campuses as biased evidence of a leftist conspiracy at work in higher education (Nocella, Best & McLaren, Forthcoming). In order to combat such alleged bias, "academic freedom" is asserted as a goal in which "both sides" of academic issues must be represented in classrooms, departments, and educational events. The result of this form of repressive tolerance (see chapter 5) is simply to impede action on matters worth acting on and to gain further ideological space for right-wing, corporate and other conservative-value agendas.[11]

It is clear, then, that despite the effects and growth of environmental education over the last few decades, it is a field that is ripe for a radical reconstruction of its literacy agenda. Again, while something like environmental education (conceived broadly) should be commended for the role it has played in helping to articulate many of the dangers and pitfalls that modern life now affords, it is also clear that it has thus far inadequately surmised the larger structural challenges now at hand and has thus tended to intervene in a manner far too facile to demand or necessitate a rupture of the status quo. What has thereby resulted is a sort of crisis of environmental education generally and, as a result, the prevailing trends in the field have recently been widely critiqued by a number of theorists and educators who have sought to highlight their limitations.

In this way, a variety of discourses and fields under monikers such as ecological education (Orr, 2004; 2002; 1992; Capra, 2002; 2000; 1996; Stone & Barlow, 2005), place-based education (Gruenewald & Smith, 2007; Haluza-DeLay, 2006), humane education (Selby, 2000; 1996; 1995; Weil, 2004), holistic education (Miller, 2007; Miller, 1991), eco-justice (Martusewicz & Edmundson, 2005; Wayne & Gruenewald, 2004; Bowers, 2001), commons-based education (Prakash & Esteva, 2008; Bowers, 2006a; 2006b; Martusewicz, 2005), transformative education (O'Sullivan, 1999; O'Sullivan, Morrell & O'Connor, 2002; O'Sullivan & Taylor, 2004; Hill & Clover, 2003), and peace education (Andrzejewski, Baltodano & Symcox, 2009; Wenden, 2004; Eisler & Miller, 2004) have been tentatively developed as either necessary counterparts to or more fit alternatives for environmental education programs generally. Most, if not all, of these approaches attempt to more robustly link forms of environmental literacy to the need for varieties of social and cultural literacy—what I define as a type of ecoliteracy. In this respect, even if these ecoliteracy frameworks move beyond sustainable development discourse in ways similar to or supportive of a critical ecopedagogy, they still arise within a growing professional trend that has also increasingly fed a call for the adoption of education for sustainable development programs around the world. Insight into the potential limitations of education for sustainable development is therefore required in order to better defend more emancipatory approaches.

From Environmental Education to Education for Sustainable Development

> Developers, Developers, Developers, Developers, Developers, Developers, Developers, Developers, Developers, Developers, Developers, Developers, Developers, Developers....Yes!
> —Steve Ballmer, CEO of Microsoft Corporation (ZDNet, 2001)[12]

In 1992, at the first Earth Summit in Rio de Janeiro, Brazil, an attempt to make a systematic policy statement about the interrelationship between humanity and the Earth was conceived of and arguably demanded. It was hoped that the document would formulate the sustainability concerns of education once and for all in both ethical and ecological (as opposed to merely technocratic and instrumentalist) terms. This document, now known as the Earth Charter (http://www.earthcharter.org), failed to emerge from Rio, however. Instead, chapter 36 of the *1992 Earth Summit Report* went on to address the issue in the following manner:

> Education is critical for promoting sustainable development and improving the capacity of the people to address environment and development issues....It is critical for achieving environmental and ethical awareness, values and attitudes, skills and behavior consistent with sustainable development and for effective public participation in decision-making. (United Nations Conference on Environment and Development, 1992, p. 2)

In 1994, the founding director of the United Nations Environment Programme and organizer of the Rio Earth Summit, Maurice Strong, along with Mikhail Gorbachev, renewed interest in the Earth Charter and received a pledge of support from the Dutch government. This led to a provisional draft of the document being attempted in 1997, with the completion, ratification, and launching of the Earth Charter Initiative at the Peace Palace in The Hague occurring on June 29, 2000. The initiative's goal was to build a "sound ethical foundation for the emerging global society and to help build a sustainable world based on respect for nature, universal human rights, economic justice, and a culture of peace."[13] While hardly a perfect set of principles, the Earth Charter's announced mission was still nothing short of revolutionary, as it attempted a bold educational reformulation of how people should maintain sustainable cultural relations with nature and between each other. It thereby cast environmental, socioeconomic, and political problems together in one light, while demanding long-term and integrated responses to our growing planetary social and ecological problems (Gruenewald, 2004).

It was hoped that the United Nations General Assembly and other governmental leaders would officially recognize and pledge to adopt the Earth Charter at the 2002 Earth Summit meetings in Johannesburg, South Africa (known as the World Summit for Sustainable Development). However, the summit proved disappointing in this and many other respects. While Kofi Annan optimistically closed the summit by announcing that $235 million worth of public–private partnerships had been achieved because of the conference, and that this put sustainable development strategies firmly on the global political map, social and environmental activists found the World Summit for Sustainable Development to be a sham for mostly the same reason.[14]

The W$$D (as its critics called it, due to its apparent pro-business agenda and bad taste in staging a posh Olympics-style event on the outskirts of the Soweto shantytowns' appalling poverty) therefore articulated a central divide between large-scale corporate and governmental technocrats and the more grassroots-based theorists, activists, and educators proper. As a result of the considerable pressure exerted by the U.S. delegates, and the additional political and economic interests of the other large states and nongovernmental organizations, the summit's concluding *Johannesburg Declaration* ultimately refused to consider ratification of the holistic, pointedly socialist in spirit, and non-anthropocentric Earth Charter educational framework (Gadotti, 2008). Instead, a Decade of Education for Sustainable Development was announced by the United Nations in 2005 and education for sustainable development was promoted as the new crucial educational field to be integrated across the disciplines and at all levels of schooling.

A leading international critic of environmental education has been Edgar González-Gaudiano (2005), who rightly charges that all-too-often the theories, policies, and discursive themes of environmental education have represented voices of the advanced capitalized nations. This results in the pressing need for environmental justice, which seeks to counteract the cultural racism inherent in mainstream sustainable (and unsustainable) development strategies, being problematically overlooked by most educational programs currently dealing with environmentalism as a set of wilderness-oriented preservationist issues (McLaren & González-Gaudiano, 1995). Therefore, by promoting an intersectional ecological concept of "human security" (p. 74), González-Gaudiano has sought to displace hegemonic ideas of national security in favor of a problem-posing pedagogy that seeks knowledge of how the environmental factors that contribute to disease, famine, unemployment, crime, social conflict, political repression, and other forms of sexual, ethnic, or religious violence can be examined as complex social and economic problems deserving of everyone's attention. In this context, he has further surmised that education for sustainable development might be used as a "floating signifier" or "interstitial tactic" capable of providing diverse groups with opportunities to produce alliances as part of the construction of a new emancipatory educational discourse (González-Gaudiano, 2005).[15] Unfortunately, however, he finds it troubling for this vision that thus far those who are not environmental educators "either appear to be uninformed or have shown no interest in the inception of a Decade that concerns their work" (p. 244).

The founding editor of the *Canadian Journal of Environmental Education* and recent co-organizer of the 5th World Congress of Environmental Education, Bob Jickling (2005), is additionally worried by the preponderance of forms of instrumentalist and deterministic education for sustainable development discourse to date. In his opinion, it is extremely worrisome that a major emerging trend within education for sustainable development is to treat education as a mere method for delivering and propagating experts' ideas about sustainable development, rather than as an opportunity to work for participatory and metacognitive engagements with students over what (if anything) sustainable development even means (Jickling & Wals, 2007).

Indeed, if this is all that is to be expected of and from education for sustainable development, then it may be concluded that it basically amounts to the latest incarnation of what the social critic Ivan Illich referred to as the prison of the "global classroom" (Illich & Verne, 1981)—an opportunity to turn ecocrisis into a rallying venture for "money, manpower, and management" (Illich, 1978). Yet, it should be pointed out that despite his serious reservations, Jickling has noted that educators are already doing good work under this moniker as well (for instance, see Sterling, 2001; Scott & Gough, 2004) and that it contains potential worthy of exploration by those concerned with educating for sustainability.[16]

Against the Third Way

Akin to González-Gaudiano and Jickling, I believe that critical ecological educators should make strategic use of the opportunities afforded by the Decade of Education for Sustainable Development (see chapter 4), but that they must refrain from becoming boosters who fail to advance rigorous critiques of its underlying political economy. To my mind, it is clear that this economy is mainly the political and economic global Third Way of so-called liberal centrists like Bill Clinton, whom the *New York Times* has referred to as the "Impresario of Philanthropy" (Dugger, 2006) because of his Clinton Global Initiative and his work on behalf of disaster relief related to the recent Asian tsunami and Hurricane Katrina.[17]

The rhetoric of this approach now champions *sustainable development* as a win-win-win for people, business, and nature, in which the following policy goals are upheld: (1) development "meets the needs of the present without compromising the ability of future generations to meet their own needs" (Brundtland, 1987) and (2) development improves "the quality of human life while living within the carrying capacity of supporting ecosystems" (Munro & Holdgate, 1991). In its tendency to deploy quasi-leftist slogans, Clintonian Third Way politics claims that it wants to put a human face to globalization and that it supports inclusive educational, medical, and civic development throughout the global south in a manner much akin to that demanded by leaders in Latin America and Africa. But if this Third Way political vision really intends to deliver greater equity, security, and quality of life to the previously disenfranchised, it is especially noteworthy that it also mandates that "existing property and market power divisions [be left] firmly off the agenda" (Porter & Craig, 2004, p. 390).

A 2000 speech by Clinton to the University of Warwick exemplifies this claim and so reveals why astute globalization critics such as Perry Anderson have characterized Thirdwayism as merely "the best ideological shell of neoliberalism today" (Anderson, 2000, p. 11). In his speech, Clinton rhetorically plugs building the necessary "consensus" to allow for the opening of previously closed markets and rule-based trade, such as that sponsored by the International Monetary Fund, in the name of a global humanitarianism, which can overcome disasters such as global warming, disease, hunger, and terrorism:

> I disagree with the anti-globalization protestors who suggest that poor countries should somehow be saved from development by keeping their doors closed to trade. I think that is a recipe for continuing their poverty, not erasing it. More open markets would give the world's poorest nations more chances to grow and prosper.
>
> Now, I know that many people don't believe that. And I know that inequality, as I said, in the last few years has increased in many nations. But the answer is not to abandon the path of expanded trade, but, instead, to do whatever is necessary to build a new consensus on trade. (Clinton, 2000)

The neoliberal market mechanism remains largely the same, then, in both Third Way social welfarism and the insanely aggressive corporatism recently favored by the Bush/Cheney administration. The only major difference between them may be the nature of the trade rules and goals issued by the governing consensus. In this, the Clinton Global Initiative is a poster child for the ideology of the majority of center-left liberals, who believe that governmental administrations can learn to legislate temperance by creating evermore opportunities for intemperate economic investment in alternative, socially responsible markets. The sustainable development vision thereby proffered is of a highly integrated world society, centered and predicated on economic trade, presided over by beneficent leaders who act in the best interests of the people (while they turn an honest profit to boot).[18] However, in this respect we might wonder if in reality this turns out to be anything other than the foxes being left in charge of the hen house.

"Sustainable development" has thus increasingly become a buzzword uttered across all political lines; one is as likely to hear it in a British Petroleum commercial as on a Pacifica radio station. As noted, the United Nations also now casts it as environmental education's heir, thereby challenging every nation to begin transforming its educational policies into a global framework for ecological and social sustainability, which can be built in relatively short order. But just what kind of sustainable development is education for sustainable development supposed to stand for? Is it consonant with alter-globalization views, or is it rather synonymous with neoliberalism in either its right or left-liberal variants?

The United Nations charges institutions (especially educational institutions) to alter their norms and practices to accord with cultural conservation strategies. But can a top-down movement for organizational change really address the fundamental failures of present institutional *technique*? The ecosocialist and founder of the German Green Party, Rudolf Bahro, noted that most institutional environmental protection "is in reality an indulgence to protect the exterministic structure," which removes concern and responsibility from people so that "the processes of learning are slowed down" (Bahro, 1994, p. 164). Does education for sustainable development amount to something radically different from this?

The next decade will ultimately decide whether education for sustainable development is little more than the latest educational fad or, worse still, turns out to be a pedagogical seduction developed by and for big business-as-usual in the name of combating social and ecological catastrophes—the educational arm of what Naomi Klein (2007) has termed *disaster capitalism*. Due to the inherent ideological contradictions currently associated with the term *sustainable development*, the Decade of Education for Sustainable Development now underway demands careful attention and analysis by critical educators in this regard. Specifically, educators will need to explain how, and if, notions of sustainability offered within this model can critically question and produce reconstructive action on the well-established social and human development models (in all of their left, center, and rightist formulations).

On the other hand, it is my belief that if education for sustainable development is utilized strategically to advance the sort of radical ecopedagogy such as for which this book will begin to lay the foundations, it could be a much-needed boost to social movements that are desperately attempting to respond to the cataclysmic challenges posed by unprecedented planetary ecocrisis. In this way, what has been heretofore known as environmental education could at last move beyond its discursive marginality by joining in solidarity with critical educators, and a real hope for an ecological and planetary society could be better sustained through the widespread deployment of transformative socioeconomic critiques and the sort of emancipatory life practices that could move beyond those programmatically offered by the culture industries and the state.

The Ecopedagogy Movement

> Eco-pedagogy is not just another pedagogy among many other pedagogies. It not only has meaning as an alternative global project concerned with nature preservation (Natural Ecology) and the impact made by human societies on the natural environment (Social Ecology), but also as a new model for sustainable civilization from the ecological point of view (Integral Ecology), which implies making changes on

economic, social, and cultural structures. Therefore, it is connected to a utopian project—one to change current human, social, and environmental relationships. Therein lies the deep meaning of eco-pedagogy....

—Angela Antunes and Moacir Gadotti (2005)

Though nascent, the international ecopedagogy movement[19] represents a profound transformation in the radical educational and political project derived from the work of Paulo Freire known as *critical pedagogy*.[20] Ecopedagogy seeks to interpolate quintessentially Freirian aims of the humanization of experience and the achievement of a just and free world with a future-oriented ecological politics that militantly opposes the globalization of neoliberalism and imperialism, on the one hand, and attempts to foment collective ecoliteracy and realize culturally relevant forms of knowledge grounded in normative concepts such as sustainability, planetarity, and biophilia, on the other. In this, it attempts to produce what Gregory Martin (2007) has theorized as a much needed "revolutionary critical pedagogy based in hope that can bridge the politics of the academy with forms of grassroots political organizing capable of achieving social and ecological transformation" (p. 349).

The ecopedagogy movement grew out of discussions first conducted around the time of the Rio Earth Summit in 1992. During the years leading up to the event, environmental themes became increasingly prominent in Brazilian circles. Then, following the Summit, a strong desire emerged among movement intellectuals to support grassroots organizations for sustainability as well as worldwide initiatives such as the Earth Charter. In 1999, the Instituto Paulo Friere under the direction of Moacir Gadotti, along with the Earth Council and UNESCO, convened the First International Symposium on the Earth Charter in the Perspective of Education, which was quickly followed by the First International Forum on Ecopedagogy. These conferences led not only to the final formation of the Earth Charter Initiative but also to key movement documents such as the Ecopedagogy Charter (Spring, 2004). Gadotti and others in the ecopedagogy movement have remained influential in advancing the Earth Charter Initiative and continue to mount ecopedagogy seminars, degree programs, workshops, and other learning opportunities through an ever-growing number of international Paulo Freire institutes.[21]

As previously noted, scholars and activists interested in furthering either environmental literacy through environmental education or variants of social and environmental ecoliteracy via education for sustainable development and its many potential subfields, have a wide number of alternatives from which to choose. However, these frameworks often ultimately derive, are centered in, or are otherwise directed from relatively privileged institutional domains based in North America, Europe, or Australia—primary representatives of the global north (Brandt, 1980). The ecopedagogy movement, by contrast, has coalesced largely within Latin America over the last two decades. Due in part to its being situated in the global south, the movement has thus provided focus and political action on the ways in which environmental degradation results from fundamental sociocultural, political, and economic inequalities.[22] As González-Gaudiano (2005) has emphasized, it is exactly these types of views and protocols that are necessary for ecoliteracy in the twenty-first century, due to their being routinely left off of northern intellectual agendas in the past. However, in a manner that moves beyond González-Gaudiano's anthropocentric, social justice–oriented approach to environmental issues, the ecopedagogy movement additionally incorporates more typically northern ecological ideas such as the intrinsic value of all species, the need to care for and live in harmony with the planet, as well as the emancipatory potential contained in human aesthetic experiences of nature.[23]

In this way, the ecopedagogy movement represents an important attempt to synthesize a key opposition within the worldwide environmental movement, one that continues to be played out in major environmental and economic policy meetings and debates. Further, as an oppositional movement with connections to grassroots political groups such as Brazil's Landless Rural Workers' Movement and alternative social institutions such as the World Social Forum, but also academic departments and divisions within the United Nations Environment Programme, the ecopedagogy movement has begun to build the extra- and intra-institutional foundations by which it can contribute meaningful ecological policy, philosophy, and curricular frameworks toward achieving its sustainability goals. Still, the ecopedagogy movement might not presently demand much interest from northern educational scholars—beyond those whose specialty is in the field of international and comparative education—save for the movement's historical relationship to the critical pedagogy of Paulo Freire.

While drawing upon a range of influences,[24] ecopedagogical theory has evolved both directly out of Freire's work and indirectly through the Latin American networks for popular education (Gutierrez & Prado, 1999; Gadotti, 2009; 2000)[25] and liberation theology (e.g., Camara, 1995; Boff,

2008; 1997) where Freire's ideas have exerted great influence. Freire himself apparently intended to issue a book on ecopedagogy, which was prevented by his death in 1997. However, in a late reflection published posthumously in *Pedagogy of Indignation*, he concluded:

> It is urgent that we assume the duty of fighting for the fundamental ethical principles, like respect for the life of human beings, the life of other animals, the life of birds, the life of rivers and forests. I do not believe in love between men and women, between human beings, if we are not able to love the world. Ecology takes on fundamental importance at the end of the century. It has to be present in any radical, critical or liberationist educational practice. For this reason, it seems to me a lamentable contradiction to engage in progressive, revolutionary discourse and have a practice which negates life. A practice which pollutes the sea, the water, the fields, devastates the forests, destroys the trees, threatens the birds and animals, does violence to the mountains, the cities, to our cultural and historical memories....(Freire, 2004, pp. 46–47)

A Critical Ecopedagogy for the North

Freire's influence upon and reinvention in the work of two generations of critical pedagogues from the United States and other advanced capitalist nations has led to his well-known reputation as being one of the greatest educational figures of modern times. Therefore, Freire's belief that today's emancipatory educational ventures must strive to combat ecocrisis means that a transformative critique of critical pedagogy as developed in northern contexts can now be made in the comparative light of the initial push for ecopedagogy in the south. This is further mandated because, despite the more recent move by some northern theorists associated with critical pedagogy to articulate or engage with ecological concerns,[26] the field of critical pedagogy has tended to remain historically silent on environmental matters. Moreover, some critics like C. A. Bowers (2003a) believe that this silence is more than accidental, and that critical pedagogical theory may not only be insufficient to fully grasp planetary ecocrisis in all its complexity, but could also unconsciously reproduce unsustainable harms in its struggle for human freedom and equity.

Affirming this idea in his own recent critique of critical pedagogy, the critical theorist of education Ilan Gur-Ze'ev (2005) has written:

> Until today, Critical Pedagogy almost completely disregarded not just the cosmopolitic aspects of ecological ethics in terms of threats to present and future life conditions of all humanity. It disregarded the fundamental philosophical and exis-

tential challenges of subject-object relations, in which "nature" is not conceived as a standing reserve either for mere human consumption or as a potential source of dangers, threats, and risks. (p. 23)

Of course, those familiar with Freire's own work will recognize that environmental themes were less than explicit in most of his writing or activities—an important point especially as he had friends and influences such as Ivan Illich, Myles Horton, Herbert Marcuse, and Erich Fromm who differed significantly from him in this respect.[27] Further, while Freire's final pedagogical reflections espoused a sort of revolutionary eco-humanism that conceived of the need to dialectically overcome the objectification of human and nonhuman natures as part of a more fully inclusive vision of liberation, one also finds therein that Freire continued to speak of humanization as an ontological vocation that stands in hard opposition to the state of nonhuman animality (Freire, 2004). This foundational humanistic dualism between the "human" and the "animal" in fact runs throughout all of Freire's work and must itself be subjected to a reconstructive ecopedagogical critique.

A crucial point is therefore raised that ecopedagogy, while drawing upon a coherent body of substantive ideas, is neither a strict doctrine nor a methodological technique that can be applied similarly in all places, all times, by all peoples. As Freire himself demonstrated with his own philosophy, pedagogies and theories evolve in their historical capacities as they meet actual challenges and reflect on their potential limitations. As a burgeoning movement, ecopedagogy is itself developing rapidly through the involvement of new individuals and groups and as political actualities on the ground change. Further, North American ecopedagogy requires reimagination in the same way that Freire demanded his own pedagogy be reinterpreted and reconstructed in order to reflect the varying cultural and historical contexts in which it was situated (Freire, 1997a, p. 308).[28]

A northern ecopedagogy should therefore begin to side and dialogue with its Latin American and related southern counterparts, at least as such positions are tentatively theorized in the Eco-pedagogy and Earth Charters (Gadotti, 2003). This means also drawing upon the emancipatory commitments and potentials of Freirian and other forms of critical pedagogy as they militate against and critique northern hegemonic forms of power such as neoliberal globalization, Machiavellian imperialism, patriarchy, systemic racism, as well as other forms of structural oppression. Lastly, a Freirian ecopedagogy also analyzes schools as practical sites for ideological struggle, but with an eye to how such struggle is connected with counterhegemonic forces outside the schools in the larger society. In other words, a northern ecopedagogy must be concerned with the larger hidden curriculum of unsustainable life and look to how social movements and a democratic public sphere are proffering vital knowledge about and against it.

The Need for Marcuse and Illich

Recently, Latin American theorists of ecopedagogy have begun to connect their work to the critical theory of Herbert Marcuse (Magelhaes, 2005; Delgado, 2005) and, to a lesser degree, other members of the Frankfurt School. As recent critical readers on Marcuse assert (Kellner, Lewis, Pierce & Cho, 2008; Abromeit & Cobb, 2004), ecological politics were an important aspect of Marcuse's revolutionary critique, and he should be considered a central theorist of the relationship between advanced capitalist society and the manifestation of ecological crisis.[29] Marcuse also taught how to overcome this crisis through the creation of revolutionary struggle and the search for new life sensibilities capable of transcending the nature/culture dichotomy that the he and other Frankfurt School members saw as a driving force behind the horrors of Western civilization. Relatedly, as Andrew Light (in Abromeit & Cobb, 2004, pp. 227–35) argues, Marcuse was an often uncited but key figure in the creation of non-anthropocentric social theory. Therefore, while both Freire and Marcuse sought through their pedagogies and politics to promote the goal of humanization, Marcuse's theory can help the ecopedagogy movement to provide a sympathetic correction of the Freirian dichotomy of the human and nonhuman.

Like Marcuse, Freire vehemently defended the pedagogical primacy of biophilia.[30] As Henry Giroux notes in his introduction to Freire's *The Politics of Education*, Freire developed a partisan view of education and praxis that "in its origins and intentions was for 'choosing life'" (Giroux, 1985, pp. xxiv–xxv). Yet, Marcuse differs from Freire in that, akin to Antonio Gramsci, he began with the primacy of the political sphere through which the necessity of education was derived—politics as education. Freire's work arguably starts with the historical given of education and strives toward a goal of political action, thereby producing a politics of education or theory of education as politics (Cohen, 1998).

For this reason, Freire's work is often tailored within critical pedagogy literature as mainly relevant to education professionals and teachers. Yet,

Marcuse offers a theory of education as a political methodology that is "more than discussion, more than teaching and learning and writing" (Kellner, 2005a, p. 85). He feels that unless and until education "goes beyond the classroom, until and unless it goes beyond the college, the school, the university, it will remain powerless. Education today must involve the mind and the body, reason and imagination, intellectual and the instinctual needs, because our entire existence has become the subject/object of politics, of social engineering" (p. 85). As a result, though a critical ecopedagogy is concerned with politicizing and problematizing the organizational milieu in which standardized ecoliteracy now occurs (or fails to occur), the manner in which ecopedagogy is first and foremost a sociopolitical movement that acts pedagogically throughout all of its varied oppositional political and cultural activities is illuminated via Marcuse's influence.

Marcuse also offers imaginative and hermeneutical "conceptual mythologies" (Kellner, 2006; 1984) that can be used to read the world in novel ways and provide openings for alternative theories and practices to the dominant exterministic order. In *Eros and Civilization* (1974), he offers the archetypal images of Orpheus and Narcissus as possible "culture-heroes" (p. 161) for a "Great Refusal" (Marcuse, 1966; 1968) of the social order. In Marcuse's view, these countercultural types exist in contradistinction to that of the Freudian Prometheus—the patriarchal representation of "toil, productivity, and progress through repression," who as "the trickster and (suffering) rebel against the gods...creates culture at the price of perpetual pain" (p. 161). Of course, Prometheus[31] is also hailed as symbolizing humanity's prophetic, historical, educative and justice-seeking aspects, and in this way he became the favorite classical mythological figure of Karl Marx. Via the Marxist reading, then, Prometheus has also come to symbolize daring deeds, ingenuity, and rebellion against the powers that be to improve human life, and in this way we can read Freirian critical pedagogy as very much a promethean movement for change.

But Marcuse's Orpheus and Narcissus make valuable ecopedagogical additions to a conceptual mythos centered on Prometheus as a figure of both good and ill.[32] Notably, Orpheus was a sort of shamanic figure who is often pictured as singing in nature and surrounded by pacified animals, while Narcissus portrays the dialectic of humanity gazing into nature and seeing the beautiful reflection of itself on new terms. Marcuse's Great Refusal, then, must be thought as intending a post-anthropocentric form of cultural work in which nature and the nonhuman are profoundly humanized, meaning that they are revealed as subjects in their own right. As Marcuse writes, through the Great Refusal, "flowers and springs and animals appear as what they are—beautiful, not only for those who regard them, but for themselves" (Marcuse, 1966, p. 166).

Another counter-reading of the Prometheus myth is offered by Ivan Illich in *Deschooling Society* (1970, pp.105–16). Illich counsels therein not for the abolishment of the Promethean instinct, but for its hegemonic displacement such that a new cultural and political age can be forged through the ideas and values of collaborative Epimethean individuals.[33] Following Marcuse, Illich revisits the Prometheus myth as a tale supporting the historical emergence of patriarchy and *Homo faber*—the progenitor of the kinds of technologies and institutions that Illich believed had drowned political hope in a global cult of expectation and social control. Versions of the myth dating back to Ancient Greece depict Prometheus as a hero whose forethought could compensate for his dim-witted brother, Epimetheus, and the destructive feminine curiosity of Epimetheus's wife, Pandora. Illich notes that prior to the establishment of patriarchy, however, Pandora was actually an ancient fertility goddess whose name meant "All Giver" and that rather than being a sexual temptation, Pandora's box was a kind of ark of sanctuary and keeper of future dreams. In marrying her, then, Epimetheus became wedded to the earth and all its gifts. Thus he represents for Illich the archetype of all those who give but do not take, who care for and treasure life (especially during times of catastrophe), and who attend to the preservation of seeds of hope in the world.

Illich was undoubtedly one of the great social and educational critics of the last few decades, a polymath who was able to bring a wide-range of learning to bear on seemingly all of the crucial issues of the day. He was intimately involved in the environmental and antinuclear movements, was a leading proponent of sustainable "post-development" (Rahnema & Bawtree, 1997) subsistence culture and the need for appropriate technologies, and championed vernacular forms of learning that took place beyond the nefarious epistemological and institutional grip of standard Western science (Prakash & Esteva, 2008). It is thus puzzling that little work, especially in educational circles, has been done on Illich altogether (Morrow & Torres, 1995, p. 232) and there is only scant scholarship that examines his theoretical relevance for understanding and solving global ecological crisis (e.g., Stuchul, Esteva & Prakash, 2005).

One possible answer to Illich's veritable disappearance from current theory has been offered by David Gabbard (1993), who surmises that Illich's gadfly politics and anarchistic sentiments have so terrified educational institutions that academics have responded by more or less collusively seeking to "write him out" of ongoing discourse, thereby rendering his work professionally illegitimate.[34] Another reason that Illich's importance as an educational philosopher may have been forgotten may ironically lie in the highly successful reception that has been given to Freire's work within critical pedagogy generally.[35] Though initially close friends, political allies, and colleagues—Illich in fact helped to free Freire from jail in 1964 and then hosted him for two summers at the Center for Intercultural Documentation (CIDOC) in Cuernavaca, Mexico, while Freire prepared his work for publication in the United States—their collaboration cooled in the ensuing decades. After Freire's *Pedagogy of the Oppressed* and Illich's *Deschooling Society* became bestsellers in the early 1970s, both became intellectual superstars and leading spokespersons for a generation of young leftist scholars and activists who sought to combat academic privilege and revolutionize campus life post–May 1968. By the late 1970s, however, Freire and Illich began to openly clash on ideological issues like the necessity of schooling, the role of *conscientization* in pedagogy, and Freire's connection to the World Council of Churches.

Though Freire and Illich ultimately remained publicly cordial and privately friendly, professionally their theoretical camps split. Critical educational theorists like Henry Giroux, Stanley Aronowitz, and Michael Apple supported Freire in the 1980s, while Illich took on the role of outsider critic and maverick, much akin to friends of his like Paul Goodman and the "home schooling" movement founders John Holt and Everett Reimer. As a result, Freire and Illich exerted influence on divergent audiences and the two were less and less seen as offering complimentary and overlapping forms of radical pedagogy. The reassertion of Illichian concerns within ecopedagogy can thereby overcome a possible historical over-reliance upon merely Freirian positions within the field of critical pedagogy. Furthermore, by dialectically conceiving of the intellectual traditions of Freire and Illich as Promethean and Epimethean collaborators, the ecopedagogy movement can achieve the sort of perspective that Illich himself counseled was necessary for the politics and culture of a new ecological age.

The Cognitive Praxis of the Ecopedagogy Movement

It must be remembered that the ecopedagogy movement is not just an abstract theory or meta-theory, untethered from a sociopolitical context. As an inclusively educational social movement trying to name, reflect upon, and act in ways that ethically accord with the vicissitudes of our current planetary ecocrisis, the movement for ecopedagogy is complex, heterogeneous, situational, both formal and informal, and a historical organizational force that is both prone to change and redefinition. Just as attempts to describe something like a "global environmental movement," or even the "American environmental movement," are hopelessly doomed to over-generalization and even reification, to speak of an "ecopedagogy movement" similarly runs the risk of violently enclosing a wide-range of different practices, ideas, and geographic struggles under a falsely singular umbrella term. It will therefore prove useful to provide a classifying framework for future work in ecopedagogy to which different groups/scholars can contribute and map themselves in relationship.

In studying differing aspects of various nations' environmentalism, social movement theorists Ron Eyerman and Andrew Jamison (1991) have helpfully pinpointed three broad dimensions, or "knowledge interests" (Habermas, 1972), that all environmentally oriented movements share in their values, work, and goals. These are, respectively, the cosmological, technological, and organizational dimensions of social change that environmental movement actors struggle to propagate (Eyerman & Jamison, 1991, pp. 70–78) throughout civic debate as well as academic and other intellectual domains of ideation. These three knowledge interests can alternatively be thought of as constituting the epistemic standpoint (Harding, 2004a) of modern environmentalism as an ecoliteracy movement.

The cosmological dimension of this standpoint speaks to the transformation in worldview assumptions that ecoliteracy can provide. According to Eyerman and Jamison this transformation represents revolutionary changes in how the dominant relationship between nature and society manifests, and its success can be measured by the degree to which a popular adoption of new paradigm ecological concepts occurs, such as happened with ideas like *ecosystem* and *dynamic balance* in previous decades (Eyerman & Jamison, 1991, p. 70). The technological dimension of environmentalism's cognitive praxis attempts to convey a winning critique of dangerous and polluting technologies, on the one hand, and the promotion of alternative, appropriate, and clean technologies developed in accordance with an ecological worldview, on the other (pp. 75–76). Finally, the organizational dimension of an ecoliteracy standpoint can be described as the principle concern that "knowledge…should serve the people" such that there is an "active dissemination of scientific information" and a "popularization of ecology and its demands for relevant interdisciplinary environmental education, merged with the identities of the other new social movements" (p. 76).

In Eyerman and Jamison's view, environmental movements (as social movements) are not simply oppositional communities but are more fully "a socially constructive force" and "a fundamental determinant of human knowledge" (p. 48). In this way, environmental movements engage pedagogically with society, with their own membership, and with other movements. They thereby generate theories, new strategic possibilities, and emergent forms of identity that can be accepted, rejected, or otherwise co-opted by dominant institutional power. This, then, is what can be called the collective *cognitive praxis* (p. 44) of disparate environmental movements—that which variously integrates and blends cosmological, technological, and organizational knowledge interests out of a plethora of movement thoughts and practices.[36] Again, these do not arise in a vacuum. Part of the development of cognitive praxis is to wage transformative campaigns on behalf of these thoughts and practices, and to attempt to march through all manner of social institutions with them, especially those overtly concerned with the function of education.

For the production of educational critique from an ecopedagogical standpoint, I thus enlist the idea of cognitive praxis as a movement intellectual in order to provide a basic structure for the further theoretical investigations of this book. In so doing, I find it neither desirable nor perhaps even possible to attempt to translate the full range of movement ideas into academic discourse. Nor is this book an attempt to be a chronicle, blueprint, or manifesto of the ecopedagogy movement and its related offshoots. Rather, in what follows, I more humbly begin to offer some foundational northern contributions to ecopedagogy as a movement concerned with the cosmological, technological, and organizational dimensions of social life, that seeks to achieve victory through its ability to:

1. provide openings for the radicalization and proliferation of ecoliteracy programs both within schools and society;

2. create liberatory opportunities for building alliances of praxis between scholars and the public (especially activists) on ecopedagogical interests; and

3. foment critical dialogue and self-reflective solidarity across the multitude of groups that make up the educational left during an extraordinary time of extremely dangerous planetary crisis.

NOTES

1. It should be noted that despite the media spectacle tethering vehicular gas mileage to global warming as a primary cause, the global livestock industry contributes far and away more global warming emissions than all forms of transportation combined and should be considered a grave environmental harm. For more on this, see the UN Food and Agriculture Organization's 2006 report, *Livestock's Long Shadow* (Steinfeld, et al., 2006). In this respect, Al Gore has himself been the subject of recent critiques by animal rights organizations like PETA and some environmental groups such as Sea Shepherd Conservation Society for leaving the demand for systemic changes in livestock and dietary practices out of his agenda to combat global climate change, in order to focus instead on eco-modernization and the creation of green technological infrastructure. It should be pointed out that he has also refused to take on the nuclear industry in this regard.

2. On the disasters and their causes, see Brown (2008); Kolbert (2006); Flannery (2006); Kunstler (2005); Diamond (2005); Posner (2004); and Rees (2003).

3. Indeed, in 2001, it was revealed at the International Standing Conference for the History of Education at the University of Birmingham, UK, that aside from one purely Australian effort (Gough, 1993), as of yet there has been no rigorous attempt to reconstruct the history of environmental education proper—it is literally a discourse without a chronicle (Wolhuter, 2001). More recent work like that of Sauvé (2005) has begun to fill this gap, however.

4. Though it must be noted that fields like outdoor education are contested terrains in which norms and boundaries can be pushed to advance progressive agendas. For instance, see Russell, Sarick & Kennelly (2003) and some of the place-based accounts in Gruenewald & Smith (2007).

5. One such reconstructive project worthy of notice is the Outdoor Empowerment program (http://www.outdoorempowerment.org).

6. We should not make environmental education into a straw man. It must be emphasized that despite the prevalent forms of environmental education and literacy that are subject to critique here, the field can be defined and analyzed to include a wide number of diverse approaches that move far beyond its problematic mainstream formulation(s) (see Sauvé, 2005). Here I argue both that most of these frameworks are not endorsed by large-scale organizations for widespread adoption and that a number of these approaches are better subsumed within the emergent field of education for sustainable development in order to contest its potential one-dimensionality and so as to highlight the ongoing normalization of environmental education as an outdoor experiential pedagogy.

7. The North American Association of Environmental Education (2000) lists four essential aspects to environmental literacy: (1) Developing inquiry, investigative, and analysis skills; (2) Acquiring knowledge of environmental processes and human systems; (3) Developing skills for understanding and addressing environmental issues; and (4) Practicing personal and civic responsibility for environmental decisions. While the third and fourth aspects respectively gesture to the possibility of a politicized version of environmental education, the lack of a specific demand for critical social thought on the part of students or for the understanding of the role of power in society, coupled with the field's traditionally "bi-partisan," approach to conflict resolution, means that the potential in this literacy agenda to foment positive ecological change through educative means is significantly undermined.

8. See http://www.mnzoo.com/animals/animals_familyfarm.asp.

9. Wikipedia (http://en.wikipedia.org/wiki/Greenwash) defines *greenwash* thusly: "Greenwash (a portmanteau of green and whitewash) is a pejorative term that environmentalists and other critics use to describe the activity of giving a positive public image to putatively environmentally unsound practices."

10. See http://www.epa.gov/enviroed/basic.html.

11. Ecopedagogy has itself come under attack by conservative educational groups such as the National Association of Scholars. For instance, see http://www.nas.org/polInitiatives.cfm?Keyword_Desc=How%20Many%20Delawares?&doc_id=303.

12. Ballmer was recently ranked by Forbes.com as the twenty-fourth wealthiest individual in the world (see http://www.forbes.com/lists/2006/10/Rank_1.html).

13. http://www.earthcharter.org/innerpg.cfm?id_page=95.

14. For coverage critical of the former Bush administration's hand in the World Summit for Sustainable Development, see the stories dated August 26 to September 6, 2002 on my weblog at http://getvegan.com/blog/blogger.php. On Annan's speech, see "Sustainable Development Summit Concludes in Johannesburg: UN Secretary-General Kofi Annan Says It's Just the Beginning" at http://www.un.org/jsummit/html/whats_new/feature_story39.htm.

15. Highlighting the ambiguity and complexity involved in distinguishing between fields like environmental education and education for sustainable development, Gray-Donald & Selby (2008) have similarly written, "Environmental education is well positioned to be a unifier, to bring together different disciplines and galvanize them into unified action" (p. 18). If environmental education is conceived very broadly, I agree with them and am arguing similarly in this book. Yet, as environmental education becomes construed more narrowly, their conclusion becomes quite untenable.

16. I follow Rolf Jucker (2002) in attempting to theorize and enact "education for sustainability" as an endeavor in critical theory that seeks transformative ecoliteracy beyond a market-based or bureaucratic sustainable development approach.

17. It is worth considering whether or not Barack Obama's ideology or policy is properly placed

18. within the spectrum of the Third Way. While his administration must still bear this out, I would argue that even if he is individually further to the left of the Bill Clintons and Tony Blairs or Gordon Browns of the world (a point that is unclear), his political vision as the president cannot itself be so. Thus the critique of the sustainable development made here should be thought applicable to our current political moment in the United States. The Obama administration could be to the right of the Third Way when it is all said and done, but it is unlikely to be left of it without the kind of public pressure that a critical ecopedagogy would work for and support.

18. While not specifically championing Third Way economics, it is remarkable how leading environmental thinkers of the present moment who *understand* that capitalism is a primary cause of planetary ecocrisis still wind up endorsing it in the variety iterated here (e.g., Speth, 2009; the Global Scenario Group, 2002). The seeming intractability of capitalist ideology among global sustainability gurus serves to bolster Slovoj Zizek's (1999) sardonic remark, "Today, we can easily imagine the extinction of the human race, but it is impossible to imagine a radical change of the social system—even if life on earth disappears, capitalism will somehow remain intact." Sustainable development must be seen, then, in at least some instances as an outcome of the systematic failure of our political imagination.

19. A growing number of texts utilize the terminology of *ecopedagogy*, without a clear relationship to the ecopedagogy movement described here. These include works by Ahlberg (1998); Jardine (2000); Petrina (2000); Yang & Hung (2004); and Payne (2005). The work of Lummis (2002) shares some sympathies, such as a critical theory approach. The earliest use of *ecopedagogy* may have been by Gronemeyer, (1987), who described it as the merging of environmentalist politics and adult education. Ironically, at the same time it was coined by Freire's friend-cum-critic Ivan Illich (1988) to describe an educational process in which educators and educands become inscribed in abstract pedagogical systems, resulting in pedagogy as an end and not a means. As used by Illich, ecopedagogy is represented by forms of education that seek the total administration of life through mandatory pedagogical experiences of systemization. As such, he believed that the movements for lifelong education and the creation of *global classrooms* (Illich & Verne, 1981) by bureaucratic educational institutions exemplified such approaches, though he was also critical of popular environmentalist pedagogy attempting to mobilize people's sentiments for *solutions* to *problems* such as global warming, hunger, and rain forest destruction. Illich's point was that such an ecopedagogy works on a problems/solutions axis that implies a global managerialism that is abhorrent to truly sustainable living in the world. This is a vastly different idea from the way the term and concept is being defined and utilized in critical education circles today, though it is potentially of great importance for the future development of the ecopedagogy movement on the whole.

20. For background on critical pedagogy, see Kincheloe (2008) and The Paulo and Nita Freire International Project for Critical Pedagogy, online at: http://freire.mcgill.ca/.

21. I was in charge of coordinating ecopedagogy initiatives for the UCLA Paulo Freire Institute from 2003 to 2005. Other institutes exist in countries such as Argentina, Portugal, India, Korea, Malta, South Africa, and Canada.

22. Infamously, the ideological divide over environmental issues was played out during the first Earth Summit in Rio de Janeiro in 1992. While representatives from the north promoted chief concerns such as habitat conservation and species preservation, representatives from the south argued that the main environmental problems affecting the planet could be traced to hemispheric economic inequalities that led the north to over-produce and consume while the south was mal-developed and being exploited by corporations for the very natural resources that northern interests argued must be preserved. The differences in values and goals between the two sides have been labeled the Green and Brown agendas, with Green issues of conservation and preservation generally favored by financially wealthier nations/regions and Brown infrastructural issues (e.g., clean water, sanitation, population health, and happiness) favored by less monetarily wealthy countries/regions.

23. This juxtaposition between the north and the south is clearly limited in that it fails to properly account for the ideas, values, and practices of the world's indigenous peoples. Indigenous perspectives often appear to integrate northern and southern agendas in key respects, but in ways that generally run parallel to and are separate from them. However, southern governance has recently shown itself more permeable to the direct incorporation of indigenous political voices. For instance, Bolivia's president is the indigenous leader, Evo Morales. In this capacity, he has been advancing a sustainability vision consonant with the ecopedagogy movement for wider audiences in Latin America (e.g., his ideas about the *rights of nature* have been likewise adopted by the Ecuadorian government) and in international policy arenas (see his United Nations speech of April 22, 2009 at http://www.boliviaun.org/cms/?p=1108).

24. The work of complexity theory, especially that offered by the French theorist, Edgar Morin, is of particular importance to Latin American ecopedagogues.

25. A listserv run by Flavio Boleiz Junior is also of central importance in coordinating work on ecopedagogy, see http://br.groups.yahoo.com/group/ecopedagogia/.

26. For instance, see Sandlin & McLaren (2009); hooks (2009); Eryaman (2009); Malott (2008); McLaren & Kincheloe (2007); Hill & Boxley (2007); McLaren & Houston (2005); Grande (2004); Gruenewald (2003); Roberts (2003); Fawcett, Bell & Russell (2002); and Mayo (2001). In the Media Education Foundation's 2006 video, *Culture, Politics & Pedagogy*, Henry Giroux also cites the tremendous challenge for critical pedagogy imposed by the grave level of planetary environmental destruction taking place. Additionally, as one can see from her prefaces for this book and Andrzejewski, Baltodano & Symcox (2009), Antonia Darder is mindful of and active on issues of planetary ecocrisis.

27. Yet, when Freire served as Sao Paulo's Minister of Education from 1989–91, he helped to implement a far-reaching curricular reorientation called the Inter Project that contained environmental justice-oriented and other ecological coursework that was thought serviceable to urban development problems and the toxicity of favela life (see O'Cadiz, Wong & Torres, 1998).

28. To this end, with Levana Saxon, I have begun organizing the Ecopedagogy Association International (http://ecopedagogy.org) that publishes *Green Theory & Praxis: The Journal of*

Ecopedagogy (http://greentheoryandpraxis.org). This association has worked in connection with various academic and activist groups interested in direct action sustainability politics such as the Institute for Critical Animal Studies (http://criticalanimalstudies.org) and Rainforest Action Network (http://action.ran.org/index.php/Ecopedagogy).

29. Jurgen Habermas also briefly notes Marcuse's importance as an ecological theorist when he writes in his "Afterword" to the *Collected Papers, Volume Two*, "Long before the Club of Rome, Marcuse fought against 'the hideous concept of progressive productivity according to which nature is there gratis in order to be exploited'" (Kellner, 2001, p. 236).

30. It is important to note that biophilia is not simply a cultural invention of the West, but can be linked to indigenous forms of traditional ecological knowledge (Cajete, 1999b) such as argued for in chapter 4.

31. Prometheus, the Greek titan whose name means *forethought*, stole the element of fire from the gods to give to humankind because his brother Epimetheus (or *afterthought*) was required to give traits to all the beings of the earth but, lacking forethought, gave them all away before he reached humanity. As a result of his theft of the divine fire, Prometheus was condemned to eternal bondage on a mountaintop where an eagle fed perpetually upon his liver.

32. Marcuse's final published writing during his lifetime was entitled "Children of Prometheus: 25 Theses on Technology and Society," in which he again reiterated how Promethean social forces have dominated nature and produced an industrially technological world of capitalism in which the repressed figure of Auschwitz is the historical possibility that drives technical progress. While he did not mention the figures of Orpheus or Narcissus, he continued to demand that a reconstruction of technological society needed to be made, not by placing artificial limits on that society, but rather by engaging inwardly and outwardly with the transvaluation of values made possible by the countercultural movements. Specifically, he concluded, "This advance towards the new is emerging today in the women's movement against patriarchal domination, which came of age socially only under capitalism; in the protests against the nuclear power industry and the destruction of nature as an ecological space that cut across all fixed class boundaries; and—in the student movement, which despite being declared dead, still lives on in struggles against the degradation of teaching and learning into activities that reproduce the system" (Marcuse, 1979, trans. Charles Reitz).

33. Fascinatingly, Illich commented that this idea was to his mind the most important of the entire book and interestingly the one that was least discussed and commented upon during his entire tenure as a public intellectual.

34. It should be noted, however, that an Illich movement has recently begun to resurface in education. For example, in the last few years a special interest group on Illich was officially formed within the American Educational Research Association. I am presently the Chair of this group and in this role have founded *The International Journal of Illich Studies* (http://ivan-illich.org/journal).

35. It should be noted that in *The Critical Pedagogy Reader* (Darder, Baltodano & Torres, 2008) Illich is included alongside Freire as a foundational figure for the field.

36. It is true that in exploring a movement's cognitive praxis, one cannot reveal the purely cosmological, technological, or organizational dimensions of its work. However, these categories can be hermeneutically useful for understanding the ways in which the texts and activities of a wide-range of groups develop common sets of understanding and hope in order to build a wider movement-oriented process for social change.

Chapter One

Cosmological Transformation as Ecopedagogy: A Critique of *Paideia* and *Humanitas*

> Philosophy, in one of its functions, is the critic of cosmologies. It is its function to harmonise, refashion, and justify divergent intuitions as to the nature of things. It has to insist on the scrutiny of the ultimate ideas, and on the retention of the whole of the evidence in shaping our cosmological scheme.
>
> —Alfred North Whitehead (1970)

Introduction

In its simplest terms, a cosmology is simply "a story of the universe and the place of the Earth and human beings in the universe at large" (Best & Kellner, 2001, p. 134). To think about cosmological shifts in society is to recognize that there exists a dominant worldview that tends to formatively gird societal ideology and people's conceptual possibilities. In the context of critiques of modern, industrialized society (e.g., Horkheimer & Adorno, 2002; White, 1996), a Western worldview has been genealogically educed that reveals—despite many discontinuities—the development of long-standing ideas over the previous millennia that have led to a form of dichotomous subjectivity and existence that is destructively symptomatic. As Steven Best and Douglas Kellner (2001) state:

> Cosmologies are integral to our self-identity, since they contextualize human existence in the broadest framework...and thereby assign meaning to daily struggles.

> Cosmologies are not always benign, however, as throughout history they have been used as the basis for establishing power and legitimating social authority. (pp. 134–35)

Over the last decade in particular, a paradigm shift in Western cosmology has begun toward what Brian Swimme and Thomas Berry (1994) refer to as "the Ecozoic era"—a time period that is witnessing the transformation of the cosmos toward a "sense of cosmogenesis" (p. 2) that is characterized by forces of "differentiation, autopoesis, and communion" (p. 71). In this way, the Ecozoic era is pictured as the unfolding of a galactic democracy and people are encouraged to imagine themselves as self-individuating elements of a larger universal community that is creatively evolving out of our "Great Work" (Berry, 1999). While such an epic story can serve to provide emancipatory meaning for a world that often seems brutally absurd and meaningless, Best and Kellner (2001) relate cautionary criticism that such a narrative reproduces opportunities for the reproduction of right-wing hegemony[1] to the degree that it fails to critique the scientific vision it utilizes:

> Cosmology…cannot be separated from history and political economy. As impressive as the new cosmologies might be, none politicize the gap between science and society, integrate social theory and technology into their coevolutionary framework…or grasp the profound political changes needed for their visions of harmony to be realized. None of the new cosmologists understand that complexity and self-organization theory are coopted by conservative, free-market thinkers, betraying the ecological thrust of the new sciences, proving once again that science can be abused unless it joins with critical social theory and radical democratic politics. Moreover, few theorists strongly integrate ethics into the heart of science, which is critical for any reconstructive program. (p. 142)[2]

In this chapter, an attempt to provide some of the missing historical and political basis of Ecozoic cosmology will be made. Specifically, the historical relationship between democratic *paideia* as practiced by the ancient Athenians, its development as Hellenistic *humanitas*, and our current ecological crisis of corporate globalization and corresponding planetary extinction will be explored. The idea that *paideia* is involved in a Western project of reified *human* literacy is proposed; and while the idea that *paideia* may serve as the foundation for a progressive pedagogy for civil democracy is explored, the development of *paideia* itself is revealed to be problematically complicit with a Western legacy of domination based upon race, class, gender, and species. The chapter ends by rejecting naïve proposals of *paideia* that would fail to apprehend the problematic character of the history of *humanitas* (i.e., the humanities), but the idea of an "ecological *paideia*" is raised as a question and possibility for future ecopedagogical exploration.

Can *Paideia* Further the Aims of a Radically Democratic Sustainability Politics?

> *Homo sapiens* has been variously described as a symbol-making animal, a tool-making animal, a social animal, a political animal, a rational animal, and a spiritual animal. Each of these characteristics has been identified as the basic element which distinguishes *Homo* from the rest of animal nature and gives him his distinctively human characteristics. It may now be that *Homo* should not only be described biologically as *Homo sapiens* but socially and culturally as *Homo educans*. It may well be that the most apt way to describe the process of man's becoming human is to say that he became a *teaching and learning* animal.
>
> —R. Freeman Butts (1973, p. 21)[3]

It is not unexpected that as people come to imagine a better and more just future that their thoughts tend to turn to the education of the young. For the children, while representing the continuance of the past, also represent the possibility that tradition is not merely static and draconian upon the present, but rather it is dynamic, democratically accessible, and interpretable. Therefore, the education of youth often comes to embody the social hope that even the most undeniable of outcomes can be trained for, grasped, redirected, and transformed into something different. It is in this sense, I believe, that the critical educator Paulo Freire spoke of learning as being both a process of *historicity* and *humanization*.

Since its birth in ancient Greece, the educational/political concept of *paideia* has played a robust role during various stages of Western development in helping to formulate the ways in which the entirety of civic life is both the subject and object of human educational activity. In this sense, *paideia* can be thought of as the West's ongoing attempt to articulate what it means to be socially *civilized* and *human* (Butts, 1973, p. 86). Emerging at the dawn of democracy in Athens two and a half millennia ago, *paideia* moved the idea of education beyond simple military preparation and the tutored construction of an aristocratic class consciousness into the domain of civic institutional interaction, where a complex of cultural skills and political literacies could be learned by the young in the name of initiating them into that overarching literacy known as *Western civilization* (pp. 85–88). To investigate the origins of education as the struggle for democracy and human potential is to arrive at *paideia*.

The question is extended, then, as to whether or not a radical ecopedagogy can now draw upon the historical underpinnings of *paideia* to provide a reinvigorated model of education for sustainable, democratic futures, such as are outlined by figures such as Morrison (1995), Fotopolous (1997), or Shiva (2005). Otherwise, is the history of *paideia*, which is also consonant with a history of Western inequality and social domination, better evoked as a *via negativa* to be criticized and overcome? After three decades of attack into the hegemony that is the theoretical bulwark represented by the phrase *Western civilization*, attacks spearheaded by waves of feminists, post-structuralists, postmodernists, and multiculturalists, to name a few, can *paideia* serve any greater purpose than to be the victim of a radical critique and dialectical sublation? Or can we reconstruct the wilderness/civilization or human/animal opposition in ways that are dialectically productive for both sides?

A Tale of Two City-States:
Athens and Its Hellenistic Reinvention, from the Cultivation of Democratic *Paideia* to the *Paideia* Cult of *Humanitas*

> The greatest work of art they [the Greeks] had to create was Man. They were the first to recognize that education means deliberately moulding human character in accordance with an ideal.
> —Werner Jaeger (1945, p. xxii)

If the question remains to be answered as to whether the concept of *paideia* retains potency toward the organization of a sustainable and democratic worldview today, it cannot be denied that the history of democracy is intimately twined to its origins in ancient Athenian *paideia*. In the fifth century B.C.E., Athens experienced what the historian of education R. Freeman Butts (1973) called a *fluorescence*, as the city-state found itself the inheritor of a political situation in which its two chief competitors, Persia and Sparta, were beaten in war and undermined by slave rebellions. Athens thus began the steady consolidation of its surrounding territories and so became the bearer of a vast new economic surplus, as well as a broad base of new citizens. However, if the Athenians demonstrated imperial geopolitical ambitions during this period of history, the anti-oligarchic domestic reforms first undertaken by Solon almost a century earlier, along with Cleisthenes's rupturing of the ancient kinship clans through the establishment of territorially based suffrage in 502 B.C.E., also provided for a then emergent opportunity to support institutions concerned with furthering unprecedented levels of democracy.[4]

It was amid this Athenian revolution in democratic participation that a reconstruction in education similarly occurred. The Greeks, who already had a long history of aristocratic education that accorded with the Homeric courtly ideal of *the heroic*, now began to reconceive of education as the *paideia* of one's total civic livelihood. Increasingly, education was not seen as something *for* society, but rather was an *end in itself*—the outcome and measure of a great society. Athens became a cosmopolis and the center of the Greek world's power, a place in which citizens navigated myriad cultural influences both foreign and domestic, and Greek society on the whole became marked by a period of rapid urbanization and social differentiation. This, combined with emergent literacies related to the popularization of the arts of reading and writing, meant that Athens in its golden age must have experienced the sort of social upheaval and disorienting cultural hybridity that we know all too well in a time of globalization and worldwide media.[5] To their credit, Athenians recognized the potentials for social transformation wrought by their situation and they instituted educational norms by which students could develop the skills, values, and democratic traits that would allow Greek culture to remain self-reflective and continuous amid all its change. In this way, whereas education had previously been the privilege of a particular class's training for a specialized culture of militarism and aristocracy, the birth of Athenian *paideia* meant that education became more "broadly 'civil'—or better 'civilizing'—in the sense that it attempted to form the citizen for a life of full participation in the wide range of activities worthy of the city" (Butts, 1973, p. 86).

The result was the mass reorganization of Greek educational activities in support of a burgeoning democratic culture. Beyond the simple inculcation of youth into preformulated expectations, Athenian *paideia* instead integrated Athenian children into the broad ideals held by Athens concerning the harmony of body, mind, spirit, and polis. The education of the Athenians thus involved all manner of physical, intellectual, aesthetic, and military exercises with the expectation that as the initiation into these various cultural domains was accomplished, the legacy of Athenian democracy would be conserved and reproduced in the speeches, acts, performances, and other creative expressions of its future citizens.

The florescent period of Athens was a time of great cultural creativity, then, and this is directly relatable to the rather liberal education of Athens' youth. As Werner Jaeger (1945) suggests in connecting Athenian *paideia* to the modern German educational tradition of *Bildung*, *paideia* was in all respects a sort of ancient *cultural studies*. More than a mere regiment of instruction, Greek *paideia* was a cultivation—a way of thinking agriculturally about society in general, such that it was hoped that the careful development of citizens' humanity would lead, not just to the fruit of great individual leaders, but to a larger expressive flourishing of Greek instincts for civilization.

In this sense, Athenian *paideia* must be interpreted as not merely the process by which the young were educated, but also as the result of that process. It was the Athenian attempt to construct direct, active political responsibility in the popular assembly as much as the creation of the great works of Greek literature and philosophical thought. The training provided by the professional educators of the day—*paedotribes*, *citharists*, *grammatists*, and civic-minded Sophists—enabled *paideia* but the result was something synergistic and more then the summative workings of the various educational parts. Just what this "more" was had to do with the relationship that the Athenians ultimately had to their own freedom and how this freedom itself related back to the system that made it possible. Thus, as represented by Pericles's Funeral Oration, *paideia* was most Athenian when its students culturally expressed the dialectical tension between valuing the collective of a democratic society, on the one hand, and the supreme achievement of that society—the liberal individual—on the other. Democracy, then, was not conceived of as an ideal for which to aspire, but the Athenians could say to one another that political freedom:

> extends also to our ordinary life. There, far from exercising a jealous surveillance over each other, we do not feel called upon to be angry with our neighbour for doing what he likes, or even indulge in those injurious looks which cannot fail to be offensive, although they inflict no positive penalty. But all this ease in our private relations does not make us lawless as citizens. (as qtd. in Bookchin, 1982, p. 130)

Their *paideia* was democratic in principle, meant to represent neither the community of Athens as a whole, nor its most celebrated individual inhabitants only. Rather, it was always in the expressed relation between the two (i.e., individual—community) that *paideia* could be found. This was the democratic city-state and its individual members bred under the ideal of *autarkeia*, "individual self-sufficiency graced by an all-roundedness of selfhood" (p. 131).

However, *paideia* in the manner just put forth lasted little more than a century. Even if we could overlook the wide disparities in economic wealth and social equality that also typified the education of Pericles's Athens (e.g., Athenian philosophy was built on top of a material foundation of slavery and coercion), still, Athenian *paideia* would appear to be little more than the ancient world's version of utopia based simply upon its chronology. It may have been wonderful in the ideal, yet it was a programmatic failure in terms of its short-lived time span.[6] As Athenian society achieved ever-greater cultural and political success it turned increasingly mercenary and brutally imperial. Likewise, social hierarchies re-emerged as predominant norms of city life and tyrannical power reconsolidated itself at the heart of state control. Almost as quickly as it began, Athenian *paideia* waned. Economic gaps widened between the various social classes and the loose federation of Greek city-states became fractured. The result was that democratic politics became evermore corrupt and oppressive within Athens proper (Butts, 1973, p. 90). Finally, as the fifth century B.C.E. closed, democracy itself was temporarily overthrown, and though it was then once more to resume for a brief time, it never again gathered the public enthusiasm that had attended it upon its first germination.

Athenian *paideia* can therefore be seen as having unsuccessfully met the pressures imposed upon it by its own form of globalization crisis—that series of cultural interactions now known as the *Hellenistic dispersal*. Just as democracy had previously come to replace local oligarchic and monarchic rule, by *paideia*'s end it was undermined by the rise of the monarchies of Philip of Macedon and his son, Alexander the Great, whose own empire finally stretched from Egypt to as far as Asia Minor. Conceptually, this same process was concretized in the philosophy of Aristotle, Plato's student and Alexander's tutor, whose mixture of aristocratic politics, scientific hierarchy, conceptual categorization, and encyclopedic breadth mirrored well the turn away from the condensation of the world into the Athenian polis toward the extension of the city-state to the rest of the world in the form of colonizing empire.

Hellenizing the Western world, Alexander brought along with his troops the very Aristotelianism that would promote *paideia* as *advanced culture*, though he failed to correspondingly propagate the previous Athenian emphasis upon democratic-process that had given rise to cultural flowering. The end result of Alexander's march was a sort of cultural revolution throughout the ancient world, with Greek armies involved in subduing and civilizing so-called "barbarous" and "inhuman" regions—first by arrow or sword, then via *paideia*:

> The most significant characteristic of the Greeks is that no group of them settled anywhere without at once establishing a school, and organized education was the most important single factor in the process of hellenization and also in the resistance to that process. (Hadas, 1959, p. 59)

Rather quickly in response to such policies and practices, the Hellenistic world began to form a far-reaching, civilized network of Greek-speaking communities oriented around Greek cultural norms. However, the lived ideal of democratic *paideia* as the full individuation of each person was steadily replaced as a goal during this time. Instead, Alexandrian elites placed an emphasis upon the high-minded imitation of what was taken to be *paideia*'s most noble accomplishments: the culture of metaphysical abstraction and the aesthetic products fashioned by an intellectual and literary sensibility (Butts, 1973, p. 107). Interested far more in achieving the clothing of high-culture, as represented by the classical literature of the past, than in educating citizens for the ethical and moral dynamics of free civic life, the Hellenistic world reconstructed *paideia* so as to meet the political needs of its ruling class. These were antidemocratic needs that were spiritually transcendent and esthetically focused, in contrast to the former Athenian emphasis on the growth of a community of relative equality among citizens.

A sort of bastardization of Athenian *paideia*, the Hellenistic age went on preserving and stylizing what it took to be the best representations of the past for nearly half of a millennia and there can be little argument that we today continue to live in the Hellenistic image and feel its affects.[7] Again, the immediate effect of the Hellenistic emphasis upon life lived as "literal homage" was that the Hellenistic world became broadly civilized in the standards of classical Athenian culture, with education centering upon the book learning and print literacy necessary to imbibe a canon of classic texts, as elites set the curricula of the newly state-controlled institutions of elementary and secondary education. Further, the Hellenistic age also erected vast systems of higher education for the specialized, aristocratic classes—from a plethora of philosophical and rhetorical research centers to the vast libraries and museums of its monarchical cities. But, lost in this immense maze of learned research, educational bureaucracy, and institutionalization of the past, was the production of knowledge for the growth of civic freedom and the realization of a better society in the future. Instead, in having become an end in itself, the Hellenistic representations of knowledge based in cultish adherence to classical forms confirmed for a resurgently powerful aristocratic class its most deeply felt hopes and fears about its own historical worthiness, even as it legitimated the aristocracy's political and economical right to rule (p. 113).

We might pause to wonder about the relationship between Athenian *paideia* and its Hellenistic transformation. I am arguing here that while the two educational projects had different cultural means and ends, with the former tending toward democratic civil service and the latter toward imperialism and the exportation of cosmopolitan culture, they are directly relatable and that Hellenistic tendencies were already at work within democratic Athens. For instance, as we have seen, even as a radical experiment in democratic *paideia* Athens never achieved anything like an inclusive democracy (Fotopolous, 1997), as it rested upon certain foundational oppressions based on slavery, race, class, gender, and species. This unresolved set of hierarchies meant that a tension existed at the very heart of the Athenian attempt at democracy. As a result, a key Athenian theme became *agonism*, and social life was constituted by values of challenge and contest. This symbolized the very violence that lay at the root of so many Athenians' perceived cultural and political freedoms. There is at work, then, a fundamental contradiction between autonomy and heteronomy, between peace and violence—and between human culture and nonhuman nature—at work in classical Athens that must be acknowledged and accounted for.

Plato himself quintessentially represents this contradiction. On the one hand, he typifies the truly exemplary individual whose great knowledge is the product of all that *paideia* offered. But, on the other, Plato also infamously utilized his education to envision and attempt to realize an antidemocratic society. Platonic thought therefore serves as the tether between autonomous and heteronomous versions of *paideia*, as his teaching was carried into the imperialism of the Hellenistic age. During the period of the Hellenistic dispersal following the fall of Athenian *paideia*, the period chronicling the conquests of Alexander through the *pax Romana*, Western civilization thrived even as changes in worldview took place and democratic communities disintegrated. The ironic result, then, was that as Hellenistic education came to define itself in relation to the historical culture of Athens, it mistook the part for the whole and so reproduced a simulated spectacle based upon classical *paideia* literature whenever it laid claim to being the true heir of the Athenian legacy (Marrou, 1964, pp. 224–25).

The ironies and contradictions of *paideia* became evermore manifest by the time of Roman *humanitas*. In his *De Oratore*, Cicero in many respects provides a re-invocation of Platonic *paideia* by stressing not only the training of the young in the *artes liberales* but also by urging their immersion into the wide-ranging, humanistic studies (*politor humanitas*) that he deemed necessary for the construction of "the good life" and its sphere of public action populated by cultivated individuals. *Paideia* as *humanitas*, then, underscored for Cicero that human excellence could only come into being if students were instructed by the broad learning of the great sciences of the past and then were properly cultured to become state leaders who could apply that learning toward the great problems of the present (Gwynn, 1966, p. 101). We should not miss the classist emphasis and bias inherent in this Roman defense of the philosopher-king, however.[8] Cicero articulated the way in which *paideia* was a sort of literacy into "becoming human" and that humanity itself was as intimately tied to the cultural heights of learned individualism as it was to the practiced maintenance of social harmony predicated upon it.

However, even this Ciceronian sense of *humanitas* aiming at the construction of the *politicos philosophos* (i.e., philosophical statesman) was not widely held during Cicero's own time. In fact, Hellenistic *humanitas* became instead even more conservative and reactionary. Under the Caesars, *humanitas* turned into a "cult of politeness" in which one's status, power, and importantly one's humanity were displayed symbolically through one's wit, high-status knowledge, and sophisticated public and private manners. Missing altogether was the Ciceronian notion of "the human" as the bearer of humane values, as well as of civilization as being the development of a civil society. In their stead, culture and political life regressed so that a "civilized man was one who was conversant with the knowledge of past civilizations, not educated to cope with the deepest crises of his own" (Butts, 1973, pp. 125–26).

Most contemporary educational theorists or historians probably do not think of *paideia* in its ancient Greek or Roman variations, though, but rather in how the concept of *paideia* was utilized in a twentieth-century American context to defend and promote a Great Books curricular program of study by the philosopher Mortimer Adler and his associates.[9] Adler defined it as follows:

> PAIDEIA (py-dee-a) from the Greek *pais, paidos:* the upbringing of the child. (Re-

lated to pedagogy and pediatrics.) In an extended sense, the equivalent of the Latin *humanitas* (from which "the humanities"), signifying the general learning that should be the possession of all human beings. (Adler, 1982, frontispiece)

Adler's insight to connect the Greek practice up with the Latin is definitely correct, though his assertion of their equivalence should be denied. Adler is also on target in identifying both *paideia* and *humanitas* as involved with the production of a liberal form of knowledge capable of differentiating and distinguishing human experience—both were interested in establishing the preeminence of a human ecology based on their principles. What Adler crucially fails to realize, and what skews his own *Paideia Proposal* (1982) in unfortunate directions, is that the Athenians and Romans never meant to confer the status of "being human" as liberally as their educational theories appear to demand.

Adler celebrates a vision of universal *humanitas* that is not born out by history. Athenian *paideia*, more progressive by comparison, still boiled down to an attempt to liberate culture from nature. It was what the philosopher Giorgio Agamben (2004) calls an *anthropological machine*, predicated on taking that which was deemed best in the world in order to refashion and inscribe it within the sensible and controlled limits of the demos. All else was either excluded outright (i.e., *barbaros*) or was otherwise made to serve the needs of the democracy by being at once excluded from power while also socially included (i.e., domesticated). *Humanitas*, as the attempt to fashion yet a second-order human nature—one identified with a deeply problematical fidelity to the images of classical representation—severed any practical relationship with the disruptive potentials afforded by democracy and became instead a technology of elitism and social hierarchy.

While there are undoubtedly worse evils than the universalization of humanistic courses of study in canonic Western literature (whether in the formats outlined by Adler, Robert Maynard Hutchins, or Earl Shorris), this continuance of the *humanitas* project errs in its conservatism even when advanced by liberals for progressive ends. Still, the point I seek to make here is not simply the multiculturalist position that we should be skeptical about the good transferred via reading the great white men of the West. My claim rather is that there are monsters lurking in the collective unconscious of any *paideia* proposal for sustainability today that require our critical attention if we want to take such a proposal seriously (see Lewis & Kahn, Forthcoming). No longer must we be committed merely to educating for the citizenship of the city-state, nor even the nation-state. In an age of unsustainable transnational capitalism, the democracy project then becomes one of *planetary citizenship*. But what is the nature of this citizenship? Are we simply extending the figure of the human in its humanist guise to the ends of the earth through a rubric of sustainable development? While it might be possible to argue that even this is more of an emancipatory political and educative vision than is presently being offered by global neoliberals (on their agenda, see Saltman, 2007), it is not clear how a global *paideia* serves to monkeywrench the anthropological machine. To my mind, planetary citizenship as imagined by the ecopedagogy movement demands the retooling of this machine as a necessary, though not clearly sufficient condition, for ecoliteracy in a time of planetary crisis.

A *Paideia* for Humanity:
History as Evolved Liberation or Entrenched Oppression?

> As the worldwide ruling class, the transnational bourgeoisie has thrust humanity into a crisis of civilization. Social life under global capitalism is increasingly dehumanizing and devoid of any ethical content. But our crisis is deeper: we face a *species crisis*. Well-known structural contradictions analysed a century ago by Marx, such as over-accumulation, under-consumption, and the tendency towards stagnation, are exacerbated by globalization, as many analysts have pointed out. However, while these "classic" contradictions cause social crisis and cultural decadence, new contradictions associated with late twentieth century capitalism—namely, the incompatibility of the reproduction of both capital *and* of nature—is leading to an ecological holocaust that threatens the survival of our species and of life itself on our planet.
> —William I. Robinson (1996)

Edmund O'Sullivan (1999), the former director of the Transformative Learning Center for the Ontario Institute for Studies in Education at the University of Toronto, has theorized that "The basic resistance to the negative fall-out of transnational globalization comes from a highly empowered civic culture that operates at the global level" (p. 256). O'Sullivan, as a promoter of what he (following the eco-theologian Thomas Berry) calls an *Ecozoic vision*, also believes that:

> A major shift took place between the "pre-modern" and "modern world" cosmologies that has had profound consequences for our thinking and actions regarding the natural world. I have indicated that the modern scientific tradition depicted nature as a non-living entity to be manipulated, controlled and exploited. (p. 105)

In the attempt to integrate C. A. Bowers's long standing call for a pedagogy of eco-justice and community, the transformative/multicultural pedagogy of someone like George Dei, and anti-oppression critical pedagogies, O'Sullivan conceives of the possibility of a newly re-invigorated ecological *paideia* that is involved in critically educating people for democratic life that will accord with what he postulates will have to be something akin to a planetary Deep Ecology experience of active caring and communion.[10]

The vision of an ecological *paideia* is compelling. But, again, in using the language of *paideia* for a new cosmological vision we must demand that it be properly historicized and politicized. We cannot afford to be social meliorists that see the evolution of Western civilization as one of the revelation of greater and greater progress only. There is also the long history that has led to an unprecedented extinction crisis and human-domination of the earth, and each may be traced back into ideas and practices of the ancient world, to the functional role of *paideia* on the nature of society, and then onward up through the Middle Ages, the Renaissance, the Enlightenment and into the Modern periods as they relate to both *paideia* and *humanitas*.[11] Such is nothing less than the history of the formation and representation of the human-as-species played back unto itself via education. Any contemporary invocation of *paideia* makes the moral demand of us, then, that we cast our critical vision back on Western civilization in the genealogical attempt to properly contextualize both the term and the current human dilemma of which it is a part.

I will not explore all of the numerous complexities involved in the assertions that I am making here, but my claim is that if we are to properly evaluate and re-fashion an image of humanity that is capable of combating and surviving the global crisis of the present moment, we must understand it as a question that emerges within the unfolding ecology that is the domain of "the human subject." Further, I want to pinpoint that a foundational element in this history of the human subject, as R. Freeman Butts's put it, was when *Homo sapiens* became *Homo educans*. This is not to say that Marx's *Homo oeconomicus*, or any of the other numerous classifications that we can confer upon humanity through an analysis of human history, should be thought as merely epiphenomenal aspects of "the educated man" or that we can (or should) reduce Western civilization to the work of human education. Rather, following the Frankfurt School (e.g., Adorno & Horkheimer's *Dialectic of Enlightenment* (1979) and Marcuse's *One Dimensional Man* (1964), I am suggesting that the various histories of the West's political, economic, intellectual and spiritual development—the story of the progress of Western culture conceived broadly—should be thought of as *the subject* of the domination of nature proper.

Paideia and *humanitas* have played significant roles in the advancement of human subjectivity, and to name an ecological *paideia* for planetary citizenship is to imagine another watershed moment in human subjectivity still. It is in this sense, then, that I would assert that we must come to a deeper understanding of *paideia*'s role in the larger history of oppression—qua human subject—and that we recognize how it supported (in both its progressive and regressive forms) the dialectic of human culture in oppositional relation to nonhuman nature. While Athenian *paideia* inscribed an entire cultural and political community, it generally failed to further embed that community within the natural world in a sustainable fashion. This dualism then became heightened during the Hellenistic age, and it is fair to assert that it has since been the dominant sociopolitical narrative that human history is the emergence of a burgeoning class of people, most previously denied human status, who then become conferred as human and so deserving of rights (only in so much as there remains a class by which to juxtapose their emancipation against). To reiterate, then, the Eurocentric history of humanism, the legacy left to us by the Hellenistic reconstruction of *paideia* in the institution of *humanitas*, achieves human rights along with the histories of speciesism, classism, sexism, and all the other histories of oppression that have led to the current entrenchment of what Riane Eisler (2000) has called *Dominator Hierarchies* (p. 4).

Interestingly, the concept of *paideia* emerges from an ideology of agriculture, with early uses of the concepts of education and cultivation as likely to reference the upbringing of plants or nonhuman animals as they were the rearing of human children.[12] Unsurprisingly, then, we can look to these agricultural beginnings for the roots of the human subject as well. Doing so, we find that at the dawn of Western civilization, "humanity" became envisioned as a sort of transitional being—partaking as much of the earthly nature of the mortal animal as that of the divine nature of the sky. This, then, is the origin for the hierarchy that posits culture as a dominant and different space from nature, and we can perceive here how a leading paradigm within Western civilization drew upon this ideological hierarchy as it began to construct a sensibility for human identity in concert with it. Hence, in early agricultural mythic-tales and cosmological narratives, like the Sumerian *Epic of Gilgamesh* or the Hebrew book of Genesis, images of the human as that which is divorced from and (at least partly) transcendent to nature, involved in urbanization processes, and semi-divine are readily apparent and central to the texts (Mason, 1998, pp. 165–72). Further, as has been widely pointed out in recent years, these tales also foster the initial codifications for the establishment of the patriarchy that would come to pervade Western social life (O'Sullivan, 1999, pp. 134–37). The overall vision of the human handed down from the cradle of civilization to the Greeks, then, was that of a dichotomous being—one ever more uneasy with its own relationship to mortal nature and so differentiating itself through identification of a transcendent immortal power, the shape of human activity related to this vision being articulated as a patriarchal, dominator culture.

By the time of Athenian *paideia*, the texts of Plato and Aristotle come to represent not only the cultural heights afforded by ancient Greek democratic educational processes, but also important ideological advancements upon the pre-Greek notion of humanity as such. Plato, as Jaeger (1945) notes, directly returned to the idea of "the divine molding" of persons out of clay when he came to theorize about the proper education of Athenian citizens (p. xxii). But Plato also went much further, and while Amelie Oksenberg Rorty (1998, p. 32) is correct in pointing out that Plato's conception of *paideia* is ultimately highly complex and evident primarily only in the whole of his work, a single instance from the *Republic* is enough to allow us to recognize the language of the human subject as Plato comfortably theorized it. Specifically, in the act of dreaming he finds humanity associated with a higher power (i.e., transcendent Reason) whose quality supersedes and subdues animal nature (i.e., the brute desires of the body):

> [The desires] are awakened in sleep when the rest of the soul, the rational, gentle and dominant part, slumbers, but the beastly and savage part, replete with food and wine, gambols and, repelling sleep, endeavors to sally forth and satisfy its own instincts. You are aware that in such case there is nothing it will not venture to undertake as being released from all sense of shame and all reason. It does not shrink from attempting to lie with a mother in fancy or with anyone else, man, god, or brute. It is ready for any foul deed of blood; it abstains from no food, and, in a word, falls short of no extreme of folly and shamelessness. (Plato, 1961, p. 798)

Thus, it was Plato's great invention to take the essence of the ancient cosmological sense of humanity's place in the world as both a demigod and the fallen steward of all things mortal, interiorize it so as to reveal a hierarchy of particular human faculties, and then reproduce this same hierarchy as a socio-political system. Paradigmatically, he thereby translates early civilization's tripartite division of god/human/animal being into a set of divides lodged within humanity proper, symbolized by the hierarchical faculties of intelligence/spirit/passion. Thereby, Plato also made humanity stand in a dialectical relationship to the world in which it was both master and slave.

Plato's student Aristotle emphasized the naturalization of the Platonic hierarchy of god/human/animal. However, he did so by forging a political vision in which the free man, under God, was handed dominion of women, children, slaves, animals and the rest of the natural world (Fouts, 1997, p. 49). Sadly, the naturalized politics of Aristotle has been used repeatedly over the ages to legitimate gender, race, class and species domination; and wherever a group of people or a nation is declared a crowning achievement of nature, Aristotle's ontological hierarchy is surely not to be far behind. Based upon either the presence or lack of what he found to be the more narrowly conceived cognitive faculties by which he defined humanity, Aristotle delimited a strict dichotomy between master and slave, which has led to highly unfortunate historical consequences for those beings that have been deemed masterable. Tellingly, Aristotle equated women with being both "unfinished" men and like the soil, a mere body whose purpose is to be the begetter of the creative seed man sows within her. Further, his justification for the institution of slavery and of the subjection of animals and other aspects of nature to the whims of those who rule was based upon his belief that all nonhuman things partake of a similar underclass:

> And it is clear that the rule of the soul over the body, and of the mind and the rational element over the passionate, is natural and expedient; whereas the equality of the two or the rule of the inferior is always hurtful. The same holds good of animals in relation to men; for tame animals have a better nature than wild, and all tame animals are better off when they are ruled by man; for then they are preserved. Again, the male is by nature superior, and the female inferior; and the one rules, and the other is ruled; this principle, of necessity, extends to all mankind.
>
> Where then there is such a difference as that between soul and body, or between men and animals (as in the case of those whose business is to use their body, and who can do nothing better), the lower sort are by nature slaves, and it is better for them as for all inferiors that they should be under the rule of a master. (Aristotle, 1943)

If the natural continuum that made up Aristotle's chain of being was conceived of as a sort of graphical plot for the history of human rights, privileged males would fall to the one side, the natural kingdom to the other, and there would be a large grey area in between comprised of beings ambiguously attempting to traverse from one side to the other. Ruling men, by their own definition—created in god's image and endowed with reason, come to represent that which is human, but they do so only in as much as they are further disembedded from their animal nature. Women, the working underclass, slaves, those of other races, all come to spend the next two millennia fighting for the rights due their "humanity" and for an equal voice in civil society with their fellow male elites. But these various histories—the histories of the struggles of race, class, and gender—have achieved liberation only at the expense of the additional underclass(es) that continue to represent that which distinguishes the nonhuman from the human. Therefore, while even Aristotle still conceived of the human as both natural and animal, the dichotomy between human and nonhuman has been strengthened and furthered considerably since.

Again, Plato and Aristotle were the epitome of Athenian *paideia* in their works and lives. But it was not until *paideia* became reconstructed as Hellenistic *humanitas* that it came to exert a major "civilizing" force upon the historical development of the West. It is through *humanitas* that we have the Hellenistic conceptual influence upon Christianity and the Church, with Augustine propounding a typically Platonic/Aristotelian view of divine human nature and of the corrupt nature of the world in which it finds itself chained. This conception would remain the official Church view throughout the Middle Ages, a time when the "ape" was defined as a failed and degraded human being, with ritual executions of these and other animals occurring alongside the hangings of criminals, Jews, and other forces of inhumanity (Morris & Morris, 1966, p. 31).

Finally, in the age of Renaissance humanism and the subsequent Enlightenment, while forces began to emerge within Western society that allowed for more people then ever before to rise up and out of the animal world of menial labor and poverty into the civility of membership within the various courts and administrations of the modern state, the ideological dichotomy between the realm of human cultural transcendence and the base state of nature only widened. Coeval with the tremendous technological advances and insights that were made during this time, early modern ideologues such as Francis Bacon called for the binding of nature into humanity's service by placing "her upon the rack" of learned scientific investigation and making of nature a slave (Spretnak, 1999, p. 54). This, when combined with the great resurgence of Hellenism that took place amongst humanists of the period, gives statements like Alberti's, "*Natura sine disciplina caeca*" (Nature without discipline is blind) a wealth of hidden meaning. Suddenly, the development of individualistic character traits through an educational system based upon the classical knowledge disciplines, the resurrection of a cultural movement in which elite learning conveyed important messages about social status, and the inscription of nature within the cult of human achievement (rather than the opposite) all emerged together as a complex nexus of values hailed as a great inheritance from the ancient world (Bantock, 1980, pp. 17–47).

Perhaps most exemplary of the early modern spirit is the figure of René Descartes, the thinker who not only helped to establish the mathematics behind the new mechanistic worldview that came to be called the Newtonian-Cartesian paradigm, but whose "Cogito ergo sum" became the slogan by which a long *humanitas*-oriented history of the centrality of human knowledge found its apotheosis. For Descartes took the implicit dualism that had haunted the history of the human subject since its first beginnings and made it powerfully explicit. Post-Descartes, the bearers of humanity, which had always defined themselves in a tenuous relationship to the natural world that they at once inhabited but felt little kinship with, now could stand legitimately separate and demonstrate their liberation from nature through their unending control.

Very much true to the roots of Hellenistic *humanitas* established over one and a half millennia earlier, Descartes identified human beings with the thinking world of subjects, superior to and unconcerned with the world inhabited by brute material bodies. Thus, with humans and animals now clearly delineated, and with the split between intelligence (*res cogitans*) and mechanism (*res extensa*) also established, it was simply a matter of logical calculation for Descartes to conclude that animals were unconscious automata, and that he could perform vivisections upon them without the use of anesthetic because he could "Kick a dog, or vivisect a dog, and it yelped not out of pain but like the spring in a clock being struck" (Fouts, 1997, p. 49).

This period of civilization was also the time in which the West witnessed the rise of a large and powerful middle class under the unfolding logic of capitalism in what amounted to a second great *globalization* of colonization and mercantilism throughout the expanding Western world. Interestingly, *paideia* itself (though in the form of *humanitas*) was once again invoked as the means by which people might better their station in life, make a "second nature" of words, image, and culture, and refine not only matter but manner (Bantock, 1980, p. 17; ff., p. 47). As such, large systems of higher education and research were established, text production boomed due to the printing press, and the growing largesse of newfound wealth arising from the exploitation of "discovered" lands and an endless series of wars allowed for the construction of a new urban, secular cultural personage. An upwardly mobile middle class erupted on the world stage, aspiring to qualities that were previously held only by the elite and globally sophisticated, and yet this also led the middle class to remove itself more fully from identifying in a beneficial way with the natural world. This, then, was the legacy of the early modern evolution of *paideia* as *humanitas*, as the educational production of the human subject, a time of the flowering of humanistic politics and the corresponding growth of the humanities curriculum.

Planetary *Paideia* as American *E Pluribus Unum*, Will It Be Gaia or Maya?

> The wisdom proper to philosophy comes from its restraint. If the latter builds up a universalizing world, art borders it with a margin of reserved beauty. Philosophers, do your work with accuracy and suffer in silence that you be treated as poets: those who are ordinarily excluded from the city. It is better that way. Build a great work where shall be found, precisely located, all things of the world, rivers, seas, constellations…but build it so beautifully that its very beauty restrains it.
> —Michel Serres (2001, p. 390)

The genealogy of *paideia* leads from the rise of humanism to its resurrection during the eighteenth, nineteenth, and twentieth centuries in the United States. During the time of the American revolution and its wake, statesmen like Thomas Jefferson and Horace Mann outlined state-funded forms of liberal education held in common by all, in which the virtues and ideals of a new democratic republic could be informed and nourished, so that the American experiment in democracy might outlast the lives of its founders (Cremin, 1980, pp. 136–39).[13] Then, during the modern period surrounding the two World Wars, social reconstructionists like John Dewey (1997) and George S. Counts (1932) challenged schools and teachers to be more than agents for the reproduction of capitalist values and status-quo ideals, instead urging them to reinsert their work within a larger civic project bent on birthing "the most humane, the most beautiful, the most majestic civilization ever fashioned by any people" (Counts, 1932, p. 35).

Yet, with its American iteration, the story of *paideia* entered a global phase and not necessarily with entirely happy results. For American *paideia*, while celebrated the world over as a system of free, democratic public education by which the lower subjected classes could move freely up the social ladder towards the full rights demanded them as meritorious human subjects, is also at least as much a propaganda line used to obscure an educational structure used to instantiate class distinctions favorable to state and economic power (Tyack, 2000, pp. 19–20). Now, when America has become the dominant world power, a vanguard producer of popular and high cultural forms, and the self-proclaimed defender of a free, democratic, human civilization, it is likewise the world's foremost polluter, a leading unsustainable producer and consumer of market goods, and the prime imperial force behind the complete instrumentalization and extinction of the natural world.

In some ways, via the traditions of the humanities and the globalization of the American socio-educational project, *paideia* is the vehicle by which huge numbers of people have become highly literate and meaningful cultural producers. But to grant it such success as this is also necessarily to recognize how it has left billions beyond the realization of the same. Even if we accept the neoliberal leadership, such as articulated previously by the Bush administration, at its word and believe that the full extension of American-led, corporate business and education into the "less cultured" regions of the globe represents a sort of final Alexandrian attempt at mass civilization, how are we to judge the results of this project if it comes at the cost of the irrational devastation of the natural planet and the further social exclusion of those beyond saving? For sure, those calling today for an ecological *paideia* of planetary citizenship mean something other than the neoliberal globalization of American values. But in accounting for the present contradictions of *paideia*, we must recognize that one path that bears its name potentially travels in this direction.

In his essay "The Individual and the Great Society," Herbert Marcuse (in Kellner, 2001) anticipates how a more oppositional and sustainable *paideia* could be formulated. Marcuse perceived an oppressively militaristic mindset behind modern science and technology and sought to keep it from dominating contemporary attempts at the creation of a free, beautiful and humane society. In this way, Marcuse called for the re-integration of science with a critical *humanitas* (pp. 74–76). He hoped thereby to invigorate the humanities with the real world questions about the domination of nature that should confront an engaged and widely informed public sphere, as well as to inject the realm of human ethics back into the hegemony represented by the natural sciences. For Marcuse, then, a new kind of humanity required the technology and learning that could be produced through the development of a new science of life (Kellner, Lewis & Pierce, 2008).[14] If we desire something like earth democracy in the face of our current global military/industrial crisis, Marcuse's thinking is therefore worthy of re-evaluation and further consideration. Still, as I have attempted to sketch, Marcuse's invocation of *paideia* as *humanitas* invokes historical contradictions that are not easily dismissed. That Marcuse's deep-seated and radical critique of education is forced, in some sense, to articulate itself around *paideia* and *humanitas* only goes to highlight how difficult it may be to escape the constraints of the past.

The challenge facing the materialization of a free sustainable society today is not just the cosmopolitan problem of how to allow for a multiplicity of (often competing) individual choices within a civic community or cultural commons, nor is it simply the challenge of how to equitably confederalize myriad common subcultural communities into an effective democratic network at the level of a worldwide human emergence.[15] Under the conditions of transnational capitalism, I believe that both of these will be required, but if these do not accompany the re-integration of the whole of humankind within the larger *oikos*[16] of our planetary nature as a whole—if the dichotomy separating the human cultural and natural kingdoms is not overcome—then the myth-making of universal civilization will continue and with it the cultural action of domination, genocide, and global ecological catastrophe as the underside of its quest for freedom.[17]

In thinking about the rise of human civilization as the differentiation of culture from nature, and so into something ideologically un-natural, Max Weber (1958) offered this stark oracle about what he took to be the jailhouse of a world dominated by bureaucratic power structures and the total disenchantment with life they breed:

> No one knows who will live in this cage in the future, or whether at the end of this present development entirely new prophets will arise, or there will be a great rebirth of old ideas and ideals, or if neither, mechanized petrification embellished with a sort of self importance. For the last stage of cultural development, it might well be truly said: 'Specialists without spirit, sensualists without heart; this nullity imagines that it has attained a level of civilization never before achieved. (p. 182)

As I write this today, highly critical of the de-politicization of the dominant culture and the dizzying heights of power obtained by those immersed in the full glories of transnational capitalism, I must admit that it is hard for me not to imagine that Weber's prophecy has come true and that we have handed the Garden of Eden over to a nullity.

But I am reminded also of the counsel of Plato, that sage voice who encapsulates certain origins of our problem and whose work, Alfred North Whitehead once remarked, made all of subsequent Western thought into but a series of footnotes. In Plato's Allegory of the Cave, Socrates tells a tale of an underground cavern in which humanity lives in chains, alone, confused, each person endlessly facing a wall upon which images are cast from the light filtering in from behind them. No one so chained has ever been outside, and no one realizes that that which they take for meaningful and real is but merely the play of shadows. Thus, Socrates wonders: What if one person broke free and was led by the light of day out of the cave and into the world? What would this person do upon seeing, not a shadow, but the Sun? What if this person then returned to the cave so as to educate his fellow prisoners, could their reaction be anything more than derisive laughter and scorn (Plato, 1961, pp. 747–51)?

It might be argued that the Socratic allegory is none other than a master's vision of the *paideia* problematic itself—one embodying the tension involved in maintaining a commitment to the development and liberation of one's peers as one also moves idealistically beyond those peers and is liberated from them in a transcendent moment of wisdom. Yet, what is most striking to me about Plato's allegory is not this evocation of the problems of the humanization process. Rather, it is the manner in which Plato connects social emancipation from humanity's troubled condition, not to a revolutionary seizure of state power and political dominion, but to the simple ascent into the light of day as brought about by a return into the natural world proper.

Could it be that the long and winding roads of Western civilization have been little more than the stories of phantoms and chains and that our true liberation lies in a direction wholly other than we have been looking at until now?[18] Those living in the new global cities, the megalopoli that are supposed to represent humanity's crowning achievements, can no longer even see the firmament at night. What would the contemporary citizens of these cities think of the madman who returned to tell them that he had not only seen the stars but that he had come from them too? Could this be the moment that serves as the educational foundation for life in a world that includes not only ecological awareness but social justice? Of the rise of an ecopedagogy movement for a new paradigm of life on earth: for earthlings?

If so, *paideia* may illuminate a future for us yet—a time in which the idea of *cosmos* is reconstructed such that new forms of local experiences can live in harmony on the planet without having to be subjected to global institutional notions of planetary harmony.[19] This would move us to realizing a political movement beyond *e pluribus unum* (see Hardt & Negri, 2004, p. 309)—the goal is not to unify the many, but to allow the many to learn to recognize what they have in common. Increasingly, it is harder and harder to know exactly what such local configurations of this political life might look like and how their complex manner might be used to limit the very structural constraints put upon ecological democracy by the forces of contemporary hybridity.[20] Further reflection will also be required that speaks to the exact nature of the curricular and the institutional manners of an ecological *paideia* in this respect. The challenge for these is to clearly distinguish themselves from the histories and political economy informing more dangerous types of authoritarian and capitalist social reproduction pedagogies.

A new way of being in (and with) the world will entail that *being social* is not reduced to an elite cult of manners, based around the dichotomies like "human" and "nonhuman/animal" or "culture" and "nature." For those of us working in education, we can take it as a first principle for the transition to ecological democracy, then, that such a world is a place in which scholars will take an interest in the natural world beyond its conscription as a resource for humanity. This includes rejecting the attempt to translate nature into a data resource for scientific measurement and management. Education will need to be more intimate and re-communed with that which has been deemed nonhuman throughout the past than a global positioning system can presently allow. This is not to say that quantity is not an issue for ecopedagogy—a movement needs numbers. As the world undergoes unprecedented mass extinction and we continue to move to a deeply entrenched market economy of transnational technocaptial, ecopedagogy should therefore aspire to become a movement for dialogue amongst various sustainability movements (e.g., traditional ecological knowledge,[21] ecosocialism, green anarchism, slow food, DIY, left biocentrism), allowing them to learn from one another and organize in a transitional alliance.

Yet, if ecopedagogy achieved a quantitative shift in democratic citizenship without fulfilling the corresponding creation of new qualities of life and consciousness, it would be a hollow victory; that is, no victory at all. As Nietzsche (1990) wrote, "To live alone one must be an animal or a god—says Aristotle. There is yet a third case: one must be both—a *philosopher*" (p. 33). A *paideia* of planetary citizenship should aim, then, for a world of philosophers (of the strong Nietzschean variety). This is the concrete pathway toward cosmological transformation for sustainability in our present historical moment, but one whose romance can be achieved only through the deepest and most critically serious consideration of the long-growing underside of the human project. As Robert Jensen (2009) writes:

> The ultimate test of our strength is whether we would be able to persevere in the quest for sustainability and justice even if we had good reasons to believe that both projects would ultimately fail. We can't know for sure, but can we live with that possibility? Can we ponder that and yet still commit ourselves to loving action toward others and the non-human world?
>
> Said differently: What if our species is an evolutionary dead end? What if those adaptations that produced our incredible evolutionary success—our ability to understand certain aspects of how the world works and manipulate that world to our short-term advantage—are the very qualities that guarantee we will destroy ourselves and possibly the world? What if that which has allowed us to dominate will be that which in the end destroys us? What if humanity's story is a dramatic tragedy in the classical sense, a tale in which the seeds of the protagonist's destruction are to be found within, and the play is the unfolding of the inevitable fall?

NOTES

1. A related criticism of Swimme and Berry is provided by Bowers (2003b), albeit from the perspective of his eco-justice framework.

2. I can attest that some of these critiques are potentially applicable to Swimme and Berry's cosmology, having studied directly under Brian Swimme as part of my doctoral work in the Philosophy, Cosmology & Consciousness program at the California Institute of Integral Studies during 1998–99. This said, the so-called *Geologian* school articulates positively with ecopedagogy in supporting an intimate and erotic relationship with the planet, the need for a post-anthropocentric human identity, and the belief that life is inherently creative. Further, I should clarify that I found Swimme to be a masterful teacher, one whom I much admire and have certainly been influenced by.

3. For a similar formulation, see Illich (1992a, pp. 113–18).

4. For a summary of Athenian politics and the difference between political and economic democracy, see Fotopolous (1995).

5. For the relationship between technology, new forms of literacy, and the development of civilization from ancient Greece onward, see Havelock (1986) and Ong (1982). For the relationship to the current cultural moment, see Kellner (2002a).

6. For an example of overlooking Athenian oppression, see the aristocratic and quasi-fascist celebration given by Jaeger (1945) that is shamelessly proclaimed on the very first pages.

7. I thank David Ulansey for helping me to understand this connection.

8. On the hypocrisies of Cicero as aristocratic humanist, see Parenti (2003).

9. Adler and his group exerted a tremendous influence on American education throughout the 1980s and 1990s.

10. See Bowers (2001) and O'Sullivan, Morrell & O'Connor (2002) in this respect.

11. For an interesting account of how our "new globalization process" should better be conceived of as the long, historical civilization process that is bound up with the history of "white privilege," see Allen (2001, pp. 467–86).

12. See *"paideia"* in Liddell & Scott (1940).

13. Also see the educational philosophy of American founder Benjamin Rush, in Provenzo (2006, pp. 58–61).

14. I take up further explorations of Marcuse's call for a "new science" in chapters 4 and 5. It would be productive to put Marcuse's view in dialogue with Thomas Berry's as articulated in work such as *The Dream of the Earth* (1988). Both see civilization as having entered a terminal phase in which techno-fetishism needs to be overcome in order to better realize common social relations predicated upon nature's subjectivity.

15. See Morrison (1995), Fotopolous (1997), Luke (1999), Shiva (2005), and LaCapra (2009) for influential theories of radical ecological democracy of a kind theorized here.

16. The Greek term means *home* and is the root of *ecology*. What our planetary home is under current historical conditions is complex and problematic. Global unsustainability means we are increasingly all "homeless" in a profound way (Deloria & Wildcat, 2001). In this sense, I support Lewis & Cho (2006) who theorize the uncanny nature of late capitalist homes and seek to imagine critical utopian places—such as the activist Julia Butterfly Hill's dwelling in the branches of the giant sequoia Luna for two years—in which late capitalist subjects such as myself might in fact "go home again."

17. This is the *Epilogue* argument put forth by Tarnas (1991, pp. 415–41). Work toward a co-constructive, posthumanist *politeia* is being done by Haraway (2003) and Latour (2004).

18. For emerging work on *paideia* in relationship to the "new cosmopolis" that is sympathetic to the critique offered here but which seeks to defend the emancipatory legacy of the West, see Kazamias (2000).

19. Importantly, in this respect, see the article by Prakash & Stuchul (2004) that calls for a cosmological multiverse of grassroots commons forms of cultural ecoliteracy in opposition to the imperial extension of the "one world" monoculture represented by mainstream Western notions of development and education.

20. For two current provocative attempts to articulate a cosmological turn on these matters, see

McKenzie, et al., (2009) and Riley-Taylor (2002).

21. Traditional ecological knowledge emphasizes an alternative cosmological relationship to place that differs significantly from mainstream social practices. See chapter 4.

Chapter Two

Technological Transformation as Ecopedagogy: Reconstructing Technoliteracy

> The great advance of electrical science in the last generation was closely associated, as effect and as cause, with the application of electric agencies to means of communication, transportation, lighting of cities and houses, and more economical production of goods. These are social ends, moreover, and if they are too closely associated with notions of private profit, it is not because of anything in them, but because they have been deflected to private uses: a fact which puts upon the school the responsibility of restoring their connection in the mind of the coming generation, with public scientific and social interests.
>
> – John Dewey (1916)

Introduction

The ongoing debate about the nature and benefits of technoliteracy is without a doubt one of the most hotly contested topics in education today. Alongside their related analyses and recommendations, the last two decades have seen a variety of state and corporate stakeholders, academic disciplinary factions, cultural interests, and social organizations ranging from the local to the global weigh in with competing definitions of *technological literacy*. Whereas utopian notions such as Marshall McLuhan's "global village" (1964), H. G. Wells's "world brain" (1938), Teilhard de Chardin's "noosphere" (1965) imagined the positive emergence of a planetary techno-ecology,[1] the contemporary situation is perhaps better characterized as the highly complex and sociopolitically stratified global culture of media spectacle[2] and the ever-developing mega-technics of a worldwide information (Castells 1996), cum technocapitalist infotainment society (Kellner, 2003a: 11–15). In other words, while contemporary information-communication technologies (ICTs) offer ecopedagogy plenty of reasons to believe that they can further the proliferation of positive values for sustainability and the organization of actual pro-ecological resistance movements to an unprecedented degree (see Kahn & Kellner, 2006; 2005), the ecological threats posed by a global media ecology whose real political and cultural economies are shielded from popular understanding and deliberation are even more greatly manifest.

But *there are both* the possibilities to use technology to promote and develop ecoliteracy widely as well as the opportunity to critique present-day technopoly (Postman, 1992) as gravely ecologically damaging. Hence, it is worthwhile for those involved in ecopedagogy to begin to ask critical questions about the types of knowledge that may be entailed by contemporary programs of technoliteracy, what sorts of practices might most greatly inform or be informed by them, as well as what institutional formations technoliteracy can best serve and be served by in kind. Further, it should be noted that despite the many divergent and conflicting views about technoliteracy that presently exist, it is only relatively recently that existing debates have begun to be challenged and informed by oppositional movements based on race, class, gender, anti-imperialism, and the ecological well-being of all. As these varying movements begin to ask their own questions about the ever-dovetailing realms of technology and the construction of a globalized culture, political realm, and economy, we may well yet see technoliteracy at once become more multiple in one sense, even as it becomes more and more singularly important for all in another.[3]

Much has been written that describes the history of the concept of "technological literacy"[4] and, as noted, a literature attempting to chart emancipatory technoliteracies has begun to emerge over the last decade.[5] In this chapter, I do not seek to reinvent the wheel of all this research or to reproduce yet another account of the same. Yet, considering that significant variance exists in the published definitions of technoliteracy, it will prove productive to begin with a survey of the meanings of the terms *technology* and *literacy* in order to more precisely conclude what sort of knowledge and skills "technoliteracy" might hail.

From this, I will summarize the broad trajectories of development in hegemonic programs of contemporary technoliteracy, beginning with their arguable origins as "computer literacy" in the U.S. *A Nation at Risk* report of

1983, through the Clinton years and the economic boom of ICTs in the 1990s, up to the more recent call for integration of technology across the curriculum and the standards-based approach of the *No Child Left Behind Act of 2001* and 2004's U.S. *National Educational Technology Plan*. Agreeing with Stephen Petrina (2000b) that the dominant trend in the United States on these matters over the last few decades has been toward the construction of a neutralized version of technoliteracy, which bolsters a neoliberal politics of ideological "competitive supremacy," I will show how this has been tacitly challenged at the international institutional level through the sustainable development vision of the United Nations' Project 2000+ (UNESCO, 1999; 1994).

In following, I will then analyze how these contestations link up with ecopedagogy's demand for educational praxis that is at once oppositional, radically democratic, and committed to sustainability. Here it seems that we must seek a reconstruction of education such that it accords with a project of multiple literacies, and I argue for a dialectical critical theory of technology that overcomes one-sided technophobic or technophilic responses and which demands modern technology's own reconstruction in favor of appropriate and liberatory forms. Finally, in closing, I think about what it will mean to reconstruct *technoliteracies*, and I propose that a major goal for ecopedagogy will be to involve people in large-scale resistance movements to actively transform mainstream understandings, policies, and practices of technoliteracy through the politicization of the hegemonic norms that currently pervade social terrains.

Technology, Literacy, Technoliteracy: Definitions

> Technological literacy is a term of little meaning and many meanings.
> —R. D. Todd (1991)

Upon first consideration, seeking a suitable definition of *technology* itself appears to be overly technical. Surely, in discussions concerning technology, it is rare indeed that people need to pause so as to ask for a clarification of the term. In a given context, if it is suggested that technology is either causing problems or alleviating them, people generally know what sort of thing is due for blame or praise.

Yet, the popular meaning of *technology* is problematically insufficient in at least two ways. First, it narrowly equivocates technological artifacts with "high-tech," such as those scientific machines used in medical and biotechnology, modern industrial apparatuses, and digital components like computers, ICTs, and other electronic media. This reductive view fails to recognize, for instance, that indigenous artifacts are themselves technologies in their own right, as well as other cultural objects that may once have represented the leading-edge of technological inventiveness during previous historical eras, such as books, hand tools, or even clothing. Secondly, popular conceptions of technology today make the additional error of construing technology as being merely object oriented, identifying it as only the sort of machined products that arise through industry. In fact, from the first, technology has always meant far more; and this is reflected in recent definitions of technology as "a seamless web or network combining artifacts, people, organizations, cultural meanings and knowledge" (Wajcman, 2004, p. 106) or that which "comprises the entire system of people and organizations, knowledge, processes, and devices that go into creating and operating technological artifacts, as well as the artifacts themselves" (Pearson & Young, 2002).

These broader definitions of technology are supported by the important insights of John Dewey. For Dewey, technology is central to humanity and girds human inquiry in its totality (Hickman, 2001). In his view, technology is evidenced in all manner of creative experience and problem solving. It should extend beyond the sciences proper, as it encompasses not only the arts and humanities, but the professions, and the practices of our everyday lives. In this account, technology is inherently political and historical and in Dewey's philosophy it is strongly tethered to notions of democracy and education, which are considered technologies that intend social progress and greater freedom for the future.

Dewey's view is hardly naïve, but it is unabashedly optimistic and hopeful that it is within the nature of humanity that people may be sufficiently educated so as to be able to understand the problems which they face and, thusly, that people can experimentally produce and deploy a wide range of technologies so as to solve those problems accordingly. While I agree strongly with the spirit of Dewey, I also recognize that the present age is potentially beset by the unprecedented problem of globalized technological oppressions in many forms (both social and environmental).

To this end, I additionally seek to highlight the insights of radical social critic and technology theorist Ivan Illich (see chapter 3). Specifically, Illich's notion of "tools" mirrors the broad humanistic understanding of technology outlined so far, while it additionally distinguishes "rationally designed devices, be they artifacts or rules, codes or operators…from other things such as food or implements, which in a given culture are not deemed to be subject to rationalization" (Illich, 1973, p. 22). Consequently, Illich polemicizes for "tools for conviviality," which are appropriate technologies mindfully rationed to work within the balances of both cultural and natural limits. In my view, technology so defined will prove useful for a twenty-first-century technoliteracy challenged to meet the demands of a sustainable and ecumenical world.

One of the great insights of Marshall McLuhan (1964) is that new media produce new environments in which people live and navigate. For instance, electricity produced entirely new urban and living spaces as well as new sciences that contributed to the development of contemporary physics and made new technologies, including the Internet. For McLuhan, a new technology of communication creates a new environment, and he has theories of the progression of stages of society and culture depending on dominant media, moving from oral culture through print culture and electronic media. New media for McLuhan require emergent literacies, and I would argue that he provides an important rationale for reconstructing education and developing the multiple technoliteracies I am discussing in this chapter in order to properly perceive, navigate, and act in an environment predicated upon the rapid evolution of industrial technology.

"Literacy" is another concept, often used by educators and policy makers, but in a variety of ways and for a broad array of purposes. In its initial form, basic literacy equated to vocational proficiency with language and numbers such that individuals could function at work and in society. Thus, even at the start of the twentieth century, literacy largely meant the ability to write one's name and decode popular print-based texts, with the additional goal of written self-expression only emerging over the following decades. Street (1984) identifies these attributes as typical of an autonomous model of literacy that is politically rightist in that it is primarily economistic, individualistic, and is driven by a deficit theory of learning. On the other hand, Street characterizes ideological models of literacy as prefiguring positive notions of collective empowerment, social context, the encoding and decoding of nonprint-based and print-based texts, as well as a progressive commitment to critical thinking-oriented skills.

In my conception, literacy is not a singular set of abilities but is multiple and comprises gaining competencies involved in effectively using socially constructed forms of communication and representation. Learning literacies requires attaining competencies in practices and in contexts that are governed by rules and conventions, and I see literacies as being necessarily socially constructed in educational and cultural practices involving various institutional discourses and pedagogies. Against the autonomous view that posits literacy as static, I see literacies as continuously evolving and shifting in response to social and cultural changes, as well as the interests of the elites who control hegemonic institutions. Further, it is a crucial part of the literacy process that people come to understand hegemonic codes as "hegemonic." Thus, my conception of literacy follows Freire and Macedo (1987) in conceiving literacy as tethered to issues of power. As they note, literacy is a cultural politics that "promotes democratic and emancipatory change" (p. viii) and it should be interpreted widely as the ability to engage in a variety of forms of problem posing and dialectical analyses of self and society.

Based on these definitions of technology and literacy it should be obvious that, holistically conceived, literacies are themselves technologies of a sort—meta-inquiry processes that serve to facilitate and regulate technological systems. In this respect, to speak of *technoliteracies* may seem inherently tautological. On the other hand, however, it also helps to highlight the constructed and potentially reconstructive nature of literacies, as well as the educative, social, and political nature of technologies. Further, more than ever, we need philosophical reflection on the ends and purposes of education and on what we are doing and trying to achieve in our educational practices and institutions. Such would be a technoliteracy in its deepest sense.

Less philosophically, I see the types of contemporary technoliteracies that can support ecopedagogy as involved with the need to comprehend and make oppositional use of proliferating high-technologies, and the political economy that drives them, toward furthering radically democratic understandings of and sustainable transformations of our lifeworlds. In a historical moment that is inexorably undergoing processes of corporate globalization and technological production, it is not possible to advocate a policy of clean hands and purity, in which people simply shield themselves from new technologies and their transnational proliferation.[6] Instead, technoliteracies must be deployed and promoted that allow for popular interventions into the ongoing (often antidemocratic) economic and technological revolutions taking place, thereby potentially deflecting these forces for progressive ends like social justice and ecological well-being.

In this, technoliteracies encompass the computer, information, critical media, and multimedia literacies presently theorized under the concept "multiliteracies" (Cope & Kalantzis, 2000; Luke 2000, 1997; Rassool, 1999; New London Group, 1996). But whereas multiliteracies theory often remains focused upon digital technologies, with an implicit thrust toward providing new media job skills for the Internet age, here I would seek to explicitly highlight the social, cultural, and ecological appropriateness of contemporary technologies and provide a critique of the emergent media economy as technocapitalist (see Best & Kellner, 2001; Kellner, 1989), while also acknowledging their emancipatory potentials. Thus, in this chapter I seek to draw upon the language of "multiple literacies" (Lonsdale & McCurry, 2004; Kellner, 2000) to augment a critical ecological theory of multiple technoliteracies as I will later expound.

Functional and Market-Based Technoliteracy: United States Models

> From being a Nation at Risk we might now be more accurately described as a Nation on the Move. As these encouraging trends develop and expand over the next decade, facilitated and supported by our ongoing investment in educational technology…we may be well on our way to a new golden age in American education.
> —U.S. Department of Education (2004)

The very fledgling Internet, then known as the ARPANET due to its development as a research project of U.S. Defense Advanced Research Projects Agency (DARPA), was still a year away when the *Phi Delta Kappan* published the following utopian call for a computer-centric technoliteracy:

> Just as books freed serious students from the tyranny of overly simple methods of oral recitation, so computers can free students from the drudgery of doing exactly similar tasks unadjusted and untailored to their individual needs. As in the case of other parts of our society, our new and wondrous technology is there for beneficial use. It is our problem to learn how to use it well. (Suppes, 1968, p. 423)

However, it was mainly not until *A Nation at Risk* (1983) that literacy in computers was popularly cited as particularly crucial for education.

The report resurrected a critique of American schools made during the Cold War era that sufficient emphases (specifically in science and technology) were lacking in curriculum for U.S. students to compete in the global marketplace of the future, as it prognosticated the coming of a high-tech "information age." Occurring in the midst of the first great boom of personal computers (PCs), *A Nation at Risk* recommended primarily for the creation of a half-year class in computer science that would:

> equip graduates to: (a) understand the computer as an information, computation, and communication device; (b) use the computer in the study of the other Basics and for personal and work-related purposes; and (c) understand the world of computers, electronics, and related technologies. (National Commission on Excellence in Education, 1983)

While *A Nation at Risk* declared that experts were then unable to classify "technological literacy" in unambiguous terms, the document clearly argues for such literacy to be understood in more functional understandings of computer (Aronowitz, 1985; Apple, 1992) and information (Plotnick, 1999) literacy. Technology, such as the computer, was to be seen for the novel skill sets it afforded, and professional discourse began to hype the "new vocationalism" in which the needs of industry were identified as educational priorities (Grubb, 1996). Surveying this development, Petrina (2000b) concludes, "By the mid-1980s in the US, technology education and technological literacy had been defined through the capitalist interests of private corporations and the state" (p. 183) and Howard Besser (1993) underscores the degree to which this period was foundational in constructing education as a marketplace.

The 1990s saw the salience and, to some degree, the consequences of such reasoning as the World Wide Web came into being and the burgeoning Internet created an electronic frontier "Dot-Com" economic boom via its commercialization amid a range of personal computing hardware and software. In the age of Microsoft and America Online, computer and information skills were indeed increasingly highly necessary. Al Gore's "data highway" of the 1970s had grown an order of magnitude to become the "information superhighway" of the Clinton presidency and the plan for a "Global Information Infrastructure" was being promoted as "a metaphor for democracy itself" (Gore, 1994). Meanwhile, hi-tech social and technological transformation took hold globally under the speculative profiteering fueled by the "new economy" (Kelly, 1998).

By the decade's end, technological literacy was clearly a challenge that could be ignored only at one's peril. Yet, in keeping with the logic of the 1980s, such literacy was again narrowly conceived in largely functional terms as "meaning computer skills and the ability to use computers and other technology to improve learning, productivity, and performance" (U.S. Department of Education, 1996). Specifically, the Department of Education located the challenge as training for the future, which should take place in schools, thereby taking the host of issues raised by the information revolution out of the public sphere proper and reducing them to standardized technical and vocational competencies for which children and youth should be trained. Further, technological literacy, conceived as "the new basic" (U.S. Department of Education, 1996) skill, became the buzz word that signified a policy program for saturating schools with computer technology as well as training for teachers and students both. Thereby, it not only guaranteed a marketplace for American ICT companies to sell their technology, but it created entirely new spheres for the extension of professional development, as teachers and administrators began to be held accountable for properly infusing computer technology into curricula.

Come the time of the Bush administration's second term, the U.S. National Education Technology Plan quoted approvingly from a high schooler who remarked, "we have technology in our blood" (U.S. Department of Education, 2004, p. 4), and the effects of two decades of debate and policy on technoliteracy was thus hailed as both a resounding technocratic success and a continuing pressure upon educational institutions to innovate up to the standards of the times.[7] Interestingly, however, the plan itself moved away from the language of technological literacy and returned to the more specific term *computer literacy* (p. 13). Still, in its overarching gesture to the *No Child Left Behind Act of 2001*, which had called for technology to be infused across the curriculum and for every student to be "technologically literate by the time the student finishes the eighth grade, regardless of the student's race, ethnicity, gender, family income, geographic location, or disability" (U. S. Congress, 2001), the United States demonstrated its ongoing commitment to delimit "technological literacy" in the functional and economistic terms of computer-based competencies.[8]

Technoliteracy for Sustainable Development: United Nations Models

Who benefits, who loses? Who pays? What are the social, environmental, personal, or other consequences of following, or not following, a particular course of action? What alternative courses of action are available? These questions are not always, and perhaps only rarely, going to yield agreed answers, but addressing them is arguably fundamental to any educational program that claims to advance technological literacy for all.

—Edgar W. Jenkins (1997)

A brief examination of the United Nations' Project 2000+: Scientific and Technological Literacy for All will illuminate how technoliteracy is being conceived of at the international level. In 1993, UNESCO and eleven major international agencies launched Project 2000+ in order to prepare citizens worldwide to understand, deliberate on, and implement strategies in their everyday lives concerning "a variety of societal problems that deal with issues such as population, health, nutrition and environment, as well as sustainable development at local, national, and international levels" (Holbrook, Mukherjee & Varma, 2000, p. 1). The project's mission underscores the degree to which the United Nations conceives of technological literacy as a social and community-building practice, as opposed to an individual economic aptitude. Further, in contradistinction to the functional computer literacy movements found in the United States context, the U.N.'s goal of "scientific and technological literacy" (STL) for all should be seen as connected to affective-order precedents such as the "public understanding of science" (Royal Society, 1985) and "science-technology-society" (Power, 1987) movements that should be considered positive forerunners of the ecopedagogy movement generally.

Though directly inspired by the social development focus of 1990s World Declaration on Education, Project 2000+ also draws in large part from the Rio Declaration on Environment and Development agreed upon at the 1992 Earth Summit (UNESCO, 1999). While the Rio Declaration itself contains ample language focused upon the economic and other developmental rights enjoyed by states, such notions of development were articulated as inseparable from the equally important goals of "environmental protection" and the conservation, protection, and restoration of "the health and integrity of the Earth's ecosystem" (United Nations, 1992). Sustainable development, to reiterate, is defined by the U.N. as "development that meets the needs of the present without compromising the ability of future generations to meet their own needs" (Brundtland, 1987), and this cannot be properly separated from radical critiques of capitalism, militarism, and other constants of our present life that structure future threats and inequality into the social system. Yet, neither can sustainable development in this formulation be separated from the ability of people everywhere to gain access and understanding of the information that can help to promote sustainability.

UNESCO does not make ICTs a centerpiece of STL projects, however. Of course, a major reason that UNESCO downplays an emphasis upon computer-related technology in its approach to technoliteracy is because the great majority of the illiterate populations it seeks to serve are to be found in the relatively poor and unmodernized regions of Latin America, Africa, and Asia, where an ICT focus would have less relevance at present. A more comprehensive reason, however, is that the United Nations has specifically adopted a nonfunctional commitment to literacy, conceiving of it as multiple literacies "which are diverse, have many dimensions, and are learned in different ways" (Lonsdale & McCurry, 2004, p. 5). STL, then, calls for understandings and deployments of appropriate technology—the simplest and most sustainable technological means that can meet a given end—as part of a commitment to literacy for social justice and human dignity.[9] This is far different than in the United States, where technoliteracy has generally been reduced to a program of skills and fluency in ICTs.

Still, it would be incorrect to conclude that the United Nations is anti-computer. In fact, the institution is strongly committed to utilizing ICTs as part of its literacy and development campaigns worldwide (Wagner & Kozma, 2003; Jegede, 2002) whenever appropriate. But as it is also conscious of the ability of new technologies to exacerbate divides between rich and poor, male and female, and north and south, the United Nations promotes "understanding of the nature of, and need for, scientific and technological literacy in relation to local culture and values" (UNESCO, 1999) and believes that scientific and technological literacy is best exhibited when it is embedded in prevailing traditions and cultures and meets people's real needs (Rassool, 1999). Consequently, while the United Nations finds that technoliteracy is a universal goal of mounting importance due to global technological transformation, STL programs require that various individuals, cultural groups, and states will formulate the questions through which they gain literacy differently and for diverse reasons (Holbrook, Mukherjee & Varma, 2000).

Expanding Technoliteracy: Toward Critical Multiple Literacies

> Technical and scientific training need not be inimical to humanistic education as long as science and technology in the revolutionary society are at the service of permanent liberation, of humanization.
>
> —Paulo Freire (1972)

As this chapter has thus far demonstrated, technoliteracy should be seen as a site of struggle, as a contested terrain used by the left, right, and center of different nations to promote their own interests, and so those interested in social and ecological justice should look to define and institute their own oppositional forms. Dominant corporate and state powers, as well as reactionary and rightist groups, have been making serious use of high-technologies to educate and so advance their agendas. In the political battles of the future, then, educators (along with citizens everywhere) will need to devise ways to produce and use these technologies to realize a critical oppositional ecopedagogy that serves the interests of the oppressed, as they aim at the democratic and sustainable reconstruction of technology, education, and society itself. Therefore, in addition to more traditional literacies such as the print literacies of reading and writing,[10] as well as other nondigital new literacies (Lankshear & Knobel, 2000), I want to argue that robustly critical forms of media, computer, and multimedia literacies need to be developed as digital subsets of a larger project of multiple technoliteracies (encompassing nondigital and digital modes) that furthers the ethical reconstruction of technology, literacy, and society in an era of technological revolution.

Critical Media Literacies

With the emergence of a global media culture, technoliteracy is arguably more important than ever, as media essentially are technologies. Recently, cultural studies and critical pedagogy have begun to teach us to recognize the ubiquity of media culture in contemporary society, the growing trends toward multicultural education, and the need for a media literacy that addresses the issue of multicultural and social difference (Hammer & Kellner, 2009; Kellner, 1998). Additionally, there is an expanding recognition that media representations help construct our images and understanding of the world and that education must meet the dual challenges of teaching media literacy in a multicultural society and of sensitizing students and publics to the inequities and injustices of a society based on gender, race, and class inequalities and discrimination (Kellner, 1995). Also, critical studies have pointed out the role of mainstream media in exacerbating or diminishing these inequalities, as well as the ways that media education and the production of alternative media can help create a healthy multiculturalism of diversity and strengthened democracy. While significant gains have been made, continual technological change means that those involved in theorizing and practicing media literacy confront some of the most serious difficulties and problems that face us as educators and citizens today.

It should be noted that media culture is itself a form of pedagogy that teaches proper and improper behavior, gender roles, values, and knowledge of the world (Macedo & Steinberg, 2007). Yet, people are often not aware that they are being educated and constructed by media culture, as its pedagogy is frequently invisible and subliminal. This situation calls for critical approaches that make us aware of how media construct meanings, influence and educate audiences, and impose their messages and values. A media-literate person, then, is skillful in analyzing media codes and conventions, able to criticize stereotypes, values, and ideologies, and competent to interpret the multiple meanings and messages generated by media texts. Thus, media literacy helps people to use media intelligently, to discriminate and evaluate media content, to critically dissect media forms, and to investigate media effects and uses.

Traditional literacy approaches attempted to "inoculate" people against the effects of media addiction and manipulation by cultivating high-cultured book literacy and by denigrating dominant forms of media and computer culture (see Postman 1992; 1985). In contrast, the media literacy movement attempts to teach students to read, analyze, and decode media texts, in a fashion parallel to the advancement of print literacy. Critical media literacy, as outlined here, goes further still in its call for the analysis of media culture as technologies of social production and struggle, thereby teaching students to be critical of media representations and discourses, as it stresses the importance of learning to use media technologies as modes of self-expression and social activism wherever appropriate (Kellner, 1995).

Developing critical media literacy and pedagogy also involves perceiving how media like film or video can also be used positively to teach a wide range of topics, like multicultural understanding and education. If, for example, multicultural education is to champion genuine diversity and expand the curriculum, it is important both for groups excluded from mainstream education to learn about their own heritage and for dominant groups to explore the experiences and voices of minority and excluded groups. Thus, media literacy can promote a more multicultural technoliteracy, conceived as understanding and engaging the heterogeneity of cultures and subcultures that constitute an increasingly global and multicultural world (Courts, 1998; Weil, 1998).

Critical media literacy not only teaches students to learn from media, to resist media manipulation, and to use media materials in constructive ways, but it is also concerned with developing skills that will help create good citizens and make them more motivated and competent participants in social life. Critical media literacy can be connected with the project of radical democracy as it is concerned to develop technologies that will enhance political mobilization and cultural participation. In this respect, critical media literacy takes a comprehensive approach that teaches critical attitudes and provides experimental use of media as technologies of social communication and change (Hammer, 2006; 1995). The technologies of communication are becoming more and more accessible to young people and ordinary citizens, and can be used to promote education, democratic self-expression, and sustainability. Technologies that could help produce the end of participatory democracy (if not life on Earth as we know it), that often transform meaningful politics into media spectacles concerned only with a battle of images and which turn spectators into cultural zombies, could also be used to help invigorate critical debate and participation within the public sphere (Giroux, 2006; Kellner & Share, 2005) and augment the struggle against ecologically catastrophic political orders.

Critical Computer Literacies

To fully understand life in a high-tech and global corporate society, people should cultivate new forms of computer literacy that involve functional knowledge of how computers are assembled and how hardware and software may be built or repaired. But critical computer literacy must also go beyond standard technical notions. In this respect, critical computer literacy involves learning how to use computer technologies to do research and gather information, to perceive computer culture as a contested terrain containing texts, spectacles, games, and interactive multimedia, as well as to interrogate the political economy, cultural bias, and environmental effects of computer-related technologies (Park & Pellow, 2004; Grossman, 2004; Plepys, 2002; Heinonen, Jokinen & Kaivo-oja, 2001; Bowers, 2000).

The emergent cybercultures can be seen as a discursive and political location in which students, teachers, and citizens can all intervene, engaging in discussion groups and collaborative research projects, creating websites, producing culture-jamming multimedia for cultural dissemination, and cultivating novel modes of social interaction and learning that can increase community in an often isolating world. Computers can thereby enable people to actively participate in the production of culture, ranging from dialogue and debate on social and ecological issues to the creation and expression of their own sustainability organizations or movements. Thus, computers and the Internet can provide opportunities for multiple voices beyond the monolingual mass media, alternative online and offline communities, and enhanced political activism (Kahn & Kellner, 2005). However, to fully take part in this counterculture requires multiple forms of technoliteracy.

For not only are accelerated skills of print literacy necessary, which are often restricted to the growing elite of students who are privileged to attend adequate and superior public and private schools, but in fact it demands a critical information literacy as well. Such literacy would require learning how to distinguish between good and what Nick Burbules & Thomas Callister (2000) identify as misinformation, malinformation, messed-up information, and mostly useless information. In this sense, information literacy is closely connected with education itself, with learning where information is archived and how it relates to the production of knowledge or critical understanding. Thus, profound questions about the relationship between power and knowledge are raised concerning the definitions of high-status and low-status knowledge, who gets to produce and valorize various modes of information, whose ideas get circulated and discussed, and whose in turn are co-opted, marginalized, or otherwise silenced altogether.

Critical Multimedia Literacies

With an ever-developing multimedia cyberculture, beyond popular spectacular film and television culture, visual literacy takes on increased importance. On the whole, computer screens are more graphic, multisensory, and interactive than conventional print fields, something that disconcerted many scholars not born of the computer generation when they were first confronted with the new environments. Icons, windows, peripherals, and the various clicking, linking, and constant interaction involved in computer-mediated hypertext dictate new competencies and a dramatic expansion of what traditionally counts as literacy.

Visuality is obviously crucial, compelling users to perceptively scrutinize visual fields, perceive and interact with icons and graphics, and use technical devices like a mouse or touchpad to access the desired material and field. But tactility is also important, as individuals must learn navigational skills of how to proceed from one field and screen to another, how to negotiate hypertexts and links, and how to move from one program to another if one operates, as most now do, in a window-based computer environment. Further, as voice and sound enter multimedia culture, literacies of the ear, speech, and especially *performance* (e.g., think of the YouTube, Second Life, and Twitter-fication of the Internet) also become part of the aesthetics and pedagogies of an expanded technoliteracy that should allow for multiple methods of learning.

Contemporary multimedia environments therefore necessitate a diversity of multisemiotic and multimodal interactions that involve interfacing with word and print material, images, graphics, as well as audio and video material (Hammer & Kellner, 2009; 2001). As technological convergence develops apace, individuals will need to combine the skills of critical media literacy with traditional print literacy and innovative forms of multiple literacies to access, navigate, and critically participate in multimediated reality.[11] Reading and interpreting print was the appropriate mode of literacy for an age in which the primary source of information was books and tabloids, while critical multimedia literacy entails reading and interpreting a plethora of discourse, images, spectacle, narratives, and the forms and genres of global media culture. Thus, technoliteracy in this conception involves the ability to respond effectively to modes of multimedia communication that include print, speech, visuality, tactility, sound, and performance within a hybrid field that combines these forms, all of which incorporate skills of interpretation and critique, agency and resistance.

Reconstructing Technoliteracy

> We are, indeed, designers of our social futures.
> —New London Group (1996)

Adequately meeting the challenge issued by the concept of technoliteracy raises questions about the design and reconstruction of technology itself. As Andrew Feenberg has long argued (1999; 1995; 1991), democratizing technology often requires its reconstruction and re-visioning by individuals and, in an ecological age, this also means seriously taking up the challenges of whether humanity intends to create sustainable designs or not (Orr, 2002). "Hackers" have redesigned digital technological systems, notably starting the largely anticapitalist Open Source and Free Software movements, and indeed much of the Internet itself has been the result of individuals contributing collective knowledge and making improvements that aid various educational, political, and cultural projects. Of course, there are corporate and technical constraints to such participation in that mainstream programs and machines seek to impose their rules and abilities upon users, but part of re-visioning technoliteracy requires the very perception and transformation of those programming limits. Technoliteracy must help teach people to become more ethical producers, even more so than consumers, and thus it can help to redesign and reconstruct technology toward making it more applicable to people's needs and not just their manufactured desires.

Crucially, alternative technoliteracies must become reflective and critical, aware of the educational, social, and political assumptions involved in the restructuring of education, technology, and society currently under way. In response to the excessive hype around new media in education, it is important to maintain a critical dimension and to actively reflect upon the nature and effects of emergent technologies and the pedagogies developed to implement and utilize them. Many academic and consumer advocates of new technologies, however, eschew critique for a more purely affirmative agenda.

For instance, after an excellent discussion of emergent modes of literacy and the need to rethink education, Gunther Kress (1997) argues that we must move from critique to design, beyond a negative deconstruction to more positive construction of high-technology. But rather than following such modern logic of either/or, critical ecopedagogues should pursue the logic of both/and, perceiving design and critique, deconstruction and reconstruction, as collaborative and mutually supplementary rather than as antithetical choices. Certainly, we need to design alternative modes of ecopedagogy and sustainability curricula for the future, as well as to provide appropriate tools for more democratic social and cultural relations in support of a planetary community, but we need also to criticize misuse, inappropriate use, over-inflated claims, as well as exclusions and oppressions involved in the introduction of ICTs into formal education and everyday life around the world. Moreover, the critical dimension is more necessary than ever as we attempt to develop contemporary approaches to technoliteracy, and to design more emancipatory, sustainable, and democratizing technologies in the context of transnational capitalism (Giroux, 2006; Suoranta & Vaden, Forthcoming). In this respect, we must be critically vigilant, always striving to practice critique and self-criticism, putting in question our assumptions, discourses, and practices about contemporary technologies, as we seek to develop multiple technoliteracies and an ecopedagogy of resistance.

In other words, people should be helped to advance the multiple technoliteracies that will allow them to understand, critique, and transform the oppressive social and cultural conditions in which they live, as they become ecoliterate, ethical, and transformative subjects as opposed to objects of technological domination and manipulation. This requires producing multiple oppositional literacies for robust critical thinking; transformative reflection; and enhancing people's capacity to engage in the production of social discourse, cultural artifacts, and political action amid a (largely corporate) technological revolution. Further, as informed and engaged subjects arise through social interactions with others, a further demand for convivial tools must come to be a part of the kinds of technoliteracy that a radical reconstruction of education now seeks to cultivate.

It cannot be stressed enough: the project of reconstructing technoliteracy must take different forms in different contexts. In almost every cultural and social situation, however, a literacy of critique should be enhanced so that citizens can name the technological and ecological system, describe and grasp the technological changes occurring as defining features of the new global order, and learn to experimentally engage in critical and oppositional practices in the interests of democratization, ecological sustainability, and progressive transformation. As part of a truly multicultural order, we need to encourage the growth and flourishing of numerous standpoints (Harding, 2004a; 2004b) on technoliteracy, looking out for and legitimizing counter-hegemonic needs, values, and understandings. Such would be to propound multiple technoliteracies "from below" as opposed to the largely functional, economistic, and technocratic technoliteracy "from above" that is favored by many industries and states. Thereby, projects for multiple technoliteracies can allow reconstructive opportunities for a better world to be forged out of the present age of unfolding technological and ecological crisis.

NOTES

1. These types of ideas are far from obsolete and there has been a continued development of them within the environmental community itself (for a leading example see the Planetwork project at: http://www.planetwork.net/background.html). For the latest articulation of how a planetary technological edifice can generate a new level of planetary literacy, see Olson & Rejeski (2007).

2. On the concept of "media spectacle" see Kellner (2005b; 2003a) that builds upon Guy Debord's notion of the "society of the spectacle," which describes a media and consumer

society organized around the production and consumption of images, commodities, and staged events. "Media spectacle" defines those phenomena of media technoculture that embody contemporary society's basic values, serve to initiate individuals into its way of life, and dramatize its controversies and struggles, as well as its modes of conflict resolution.

3. The idea that different forms of knowledge (e.g., different types of questions which in turn beget different answers) are produced as an oppressed group begins to achieve a collective identity vis-à-vis the social, cultural, and political issues of the day is a central insight of the critical theory known as *feminist standpoint theory* (Harding, 2004a). It can be argued that this idea girds critical theory in general, and a radical formulation can be seen in Marcuse (1965), as well as in the works of Marx and Engels proper as Sandra Harding points out.

4. For instance, see Petrina (2000a); Selfe (1999); Jenkins (1997); Waetjen (1993); Lewis & Gagel (1992); Dyrenfurth (1991); Todd (1991); Hayden (1989).

5. See Kellner (2004; 2003c; 1998); Lankshear & Snyder (2000); Petrina (2000a); Luke (1997); Bromley & Apple, (1998); Ó Tuathail & McCormack (1999); Burbules & Callister (1996); McLaren, Hammer, Sholle & Reilly (1995).

6. Yet, stressing the social and cultural specificity of technologies, neither am I calling for the universal adoption of high technologies, nor do I link them essentially to progress as necessary stages of development. On the other hand, I urge caution against technophobic attitudes, as I favor a dialectical view of technology and society.

7. A definition of *technocracy* is offered by Kovel (1983, p. 9) as being the social order where "the logic of the machine settles into the spirit of the master. There it dresses itself up as 'value-free' technical reasoning."

8. In 2002, the International Technology Education Association issued its *Standards for Technological Literacy: Content for the Study of Technology*, which intends to be definitive for the field. To be fair, at least eight of its twenty standards evoke the possibility of affective components that move beyond the functional, market-based approaches chronicled here. However, as Petrina (2000b, p. 186) notes, the Director of the Technology for All Americans project involved in creating the standards declared that they were "the vital link to enhance America's global competitiveness in the future" and so their vocational and economic concerns must be considered central.

9. In an educational context, one interested in earlier versions of STL might look to the 1960s work of figures such as Ivan Illich and Paulo Freire (see chapter 3), who offered a critique of modern developmental strategies in the Third World and called for appropriate and democratic technological change in its place.

10. I resist that technoliteracy outmodes print literacy. Indeed, in the emergent information-communication technology environment, traditional print literacy takes on increasing importance in the computer-mediated cyberworld as people need to critically scrutinize tremendous amounts of information, putting increasing emphasis on developing reading and writing abilities. Theories of secondary illiteracy, in which new media modes contribute to the complete or partial loss of existing print literacy skills due to lack of practice, demonstrates

that new technologies cannot be counted upon to deliver print literacy of their own accord.

11. To critically participate in such a reality does not mean serving as its booster or even acquiescing to its adoption in one's life. The point here is that even someone living nondigitally in relative simplicity participates increasingly in a society and world that are moving in contrary directions. My supposition here is that the refusal represented by someone having gone back to the land is augmented by her/his being knowledgeable about what is being rejected and why. This transforms place from a life formed through naive inhabitation to one based in "decolonization and reinhabitation" (Gruenewald, 2003) of the planetary commons.

Chapter Three

The Technopolitics of Paulo Freire and Ivan Illich: For a Collaborative Ecopedagogy

THIS MACHINE KILLS FASCISTS
—Words inscribed on Woody Guthrie's guitar.

THIS MACHINE SURROUNDS HATE AND FORCES IT TO SURRENDER
—Words inscribed on Pete Seeger's banjo.[1]

Introduction

In her essay "The Social Importance of the Modern School," Emma Goldman (1912) considers the importance of history as a subject of education, noting that schools must "help to develop an appreciation in the child of the struggle of past generations for progress and liberty, and thereby develop a respect for every truth that aims to emancipate the human race." With this in mind, this chapter seeks to interrogate the legacy of radical ecopedagogues like Paulo Freire and Ivan Illich and inquire whether their struggles still live for the students of standardized curricular capitalism, whose schools are littered with corporate advertising and products, and who are themselves either tracked into broken-down buildings lacking adequate textbooks and materials or into a cutthroat competition for admissions' placement that begins with preschool and continues on through college and one's professional career.

Sadly, schools today are not regularly engaged by the emancipatory pedagogies and social movements sparked by the work of these two great mentors, perhaps the late twentieth century's most important figures in the field of education due to their wide-ranging and perceptive theories linking politics and culture, capitalist economics, and human ethics to a rigorous critique of schooling. Today, as schools cuddle up to business and replace programs for literacy with a profit-friendly "computer literacy" (Aronowitz, 1985, p. 13), steadily moving computers from the production line to "the center of the classroom" (Apple, 1992), those who currently theorize and practice education will find Freire and Illich's philosophies of education extremely relevant to the wide range of questions that the current proliferation of technology produces for pedagogy.

Routinely, culture everywhere is becoming saturated with media, in which many aspects of myriad people's lives are mediated by technology (Stone, 2001). Technologized media themselves now constitute Western culture through and through and they have become "the primary vehicle for the distribution and dissemination of culture" (Kellner, 1995, p. 35). Thus, as the sociologist Manuel Castells (1999) has noted, "Politics that does not exist in the media…simply does not exist in today's democratic politics" (p. 61). While the North American followers of Paulo Freire continue to oppose rightist mainstream educational technology policies and practices through the discourses of *critical pedagogy* and *critical media literacy*, it is surprising, then, that few works therein deal at length with Freire's own pedagogical relationship to new technologies.

More recently, neo-Illichians[2] like John Ohlinger (1995); C. A. Bowers (2000); Dana Stuchul, Gustavo Esteva and Madhu Suri Prakash (2005) have attempted to challenge Freirian critical pedagogy's iconic status in leftist educational circles by producing strong (sometimes ad hominem[3]) critiques of Paulo Freire and those he has influenced in favor of postcolonial forms of cultural ecoliteracy. However, these interventions have so far been met with little extended debate or rebuttal from both mainstream and critical educators. With the death of Freire in 1997, and Illich in 2002, the opportunity was sadly lost for each to break bread once again, jointly comment upon their important points of agreement and disagreement, and potentially reconstruct what are arguably two of the strongest radical traditions vis-à-vis education and technology.

For this reason, it is an important component of an ecopedagogy which seeks to build on the work of Freire and Illich that a less polemical and more dialectical critique is produced in which both the positives and negatives of Freire's and Illich's theories are contextualized by present-day needs, even as the two theorists are themselves compared and contrasted for affinities and differences.[4] In this chapter, I therefore will undertake what Douglas Kellner (1995) calls a *diagnostic critique*, a dialectics of the present that "uses history to read texts and texts to read history," with the end goal of grasping alternative pedagogical practices and utopian yearnings for a reconstruction of education in the future, such that criticalists will be challenged to develop pedagogies and political movements that address these challenges, while propounding radical critiques of education such as previously offered by Freire and Illich.[5]

Against one-sided critiques of present educational technology that are overly technophilic or technophobic, this chapter seeks to understand the present moment in education and society as marked by "objective ambiguity" (Marcuse, 1964). That is, reality should be seen as complex and contested by a variety of forces, rich with alternatives that are immediately present and yet ideologically, normatively, or otherwise blocked from achieving full realization in their service to society (Marcuse, 1972c, p. 13). It is therefore the utopian challenge to radicalize social practices and institutions through the application of new diagnostic critical theories and alternative pedagogies such that oppressive cultural and political features are negated, even as liberatory tendencies within everyday life are articulated and reaffirmed.

Notably, this process has been conceptualized as "reconstruction" by progressive educators like John Dewey (1897) and revolutionaries like Antonio Gramsci, who importantly noted that "every crisis is also a moment of reconstruction" in which "the normal functioning of the old economic, social, cultural order, provides the opportunity to reorganize it in new ways" (Hall, 1987).[6] To speak of technology, sustainability politics and the reconstruction of education, then, is to historicize and critically challenge current trends in education toward using the tools at hand to create further openings for transformative praxis on behalf of planetary emancipation.

The Politics of Information, Infotainment and Technocapital

> It's peculiar and unnerving in a way to see so many young people walking around with cell phones and iPods in their ears and so wrapped up in media and video games....It's a shame to see them so tuned out to real life. Of course they are free to

do that, as if that's got anything to do with freedom. The cost of liberty is high, and young people should understand that before they start spending their lives with all those gadgets.

—Bob Dylan, quoted in Brinkley (2009)

Humanity begins the twenty-first century by undergoing one of the most, if not the most, dramatic technological revolutions in history. As it is centered on computer, information, communication, and multimedia technologies, the resulting product of this revolution is often hailed as the beginning of a *network* or *information society* (Castells, 1999; 1996; Kellner, 2002b). In the hands of its many boosters, the information society has often been represented as a sort of cyber-ecumene capable of bridging differences, weaving communion, and welcoming underdeveloped regions into a form of *global village* political economy. But through the information society's impetus toward modernization and development practices, traditional forms of social organization, culture and politics are routinely being outmoded, imploded into and hybridized with novel cultural and political modes to create a highly mediated realm of *technocapitalism* (Kellner, 2000; 1989; Best & Kellner, 2001). In this respect, then, it is now clear that the digitized "one world" (Cosgrove, 2001) of harmonious planetary communication brought about by the exchange of information is in many ways a myth that cloaks the seductive inequalities of what is better characterized as an *infotainment society* (Kellner, 2003a), a globally networked economy driven by corporate and imperial forces of science, technology, and a new Internet technocultural complex.[7]

Over the last few decades, the culture industries beholden to technocapital have multiplied media spectacles throughout all manner of colonized public spheres, and spectacle itself is becoming one of the organizing principles of the economy, polity, society, and everyday life (Kellner, 2003a). The Internet-based economy deploys spectacle as a means of promotion, reproduction, and the circulation and selling of commodities. Media culture itself proliferates evermore technologically sophisticated spectacles to seize audiences and increase their power and profit. The forms of entertainment permeate news and information, and a tabloidized infotainment culture is increasingly popular. New multimedia that synthesize forms of radio, film, TV news and entertainment, and the mushrooming domain of cyberspace, become spectacles of technoculture, generating expanding sites of information and entertainment, while intensifying the spectacle form of media culture.

In the United States, the nation and culture of megaspectacle, schools have been forced to transform under the pressures wrought by ubiquitous media, technoculture, and a computer industry that seeks to place a computer in every child's hands (Trend, 2001). A relatively recent government study, *A Nation Online: How Americans Are Expanding Their Use of the Internet* (National Telecommunications & Information Administration, 2002), reveals that 90 percent of children between the ages of five and seventeen (forty-eight million) operate computers and that Internet use is increasing for people regardless of income, education, age, race, ethnicity, or gender. Additionally, the United States Department of Education cites figures that as of 2005 "nearly 100 percent of all public schools in the United States had access to the Internet," almost all had broadband connectivity, and that there was a 3.8:1 ratio of students to computers with Internet access in the public schools.[8] However, despite trends charting an increase of utilization by every demographic, Internet access in the United States remains largely stratified along lines of race, class, and level of educational attainment (Lenhart, et. al., 2003). Schools thus serve as the primary places in which all manner of youth might have the ability to interact with the global Internet, develop creative and technical technoliteracy skills, and so acquire the necessary cultural capital to understand and survive in an infotainment economy.

The critical educator Antonia Darder is undoubtedly correct when she calls attention to the fact that wealthy schools and districts often have greater access to computer technology and Internet access, and that the minority cultures that tend to comprise poorer schools and districts are placed in a role of having always to compete on an unequal playing field (Darder, 2002). However, this critique should be placed in the context of the opportunities for student and community agency that can also arise from newly infused technology in schools and community centers. Further, we should pay attention to the ways in which poor school districts sometimes capitalize upon their underserved and minority status to apply for and win state, federal, and corporate technology grants. For example, the Lennox School District (in Los Angeles County), a district in which median household incomes are below the national average, unemployment is above the national average, and Spanish is the primary language spoken among a 97 percent Latino/Chicano population, has been awarded hundreds of thousands of dollars in development grants through applications to the state and federal government. Further, Lennox "teamed" with Apple Corporation as a partner in the company's PowerSchool Information System initiative that wired the district in order to provide a system in which teachers, students, administrators, and parents can all have real-time access to information about student, class, and school progress.[9]

I cite this example to point out the need for critical educators to integrate their theories and practices with the often contradictory and multifaceted realities at work today in the lives of oppressed peoples. Lennox's technology initiative has unquestionably transformed its schools, providing a level of technological infusion unmatched by even the wealthy Beverly Hills School District to its north, and it has used its status as a poor, minority district toward achieving this end. Yet, the question remains as to how this technology is affecting the lives of students and families in the area for both good and ill. That Lennox's PowerSchool seeks to more closely monitor students' work and lives might trigger cause for alarm, as a post-Columbine (and post-Cho) paradigm in education points toward the use of information technologies and the psychological profiling of students to create sophisticated tools of administrative surveillance and discipline that function freely under the general claim of "security" (Lewis, 2003). As schools in Lennox have historically suffered gang-related violence, resulting in policy emphases upon disciplinary focus and increased safety measures, suspicion and a closer examination of the school district's corporate-fed information system are warranted.

In a nonformal educational context, my work with Douglas Kellner (Kahn & Kellner, 2008; 2007; 2006; 2005) has demonstrated the manner in which changes in global society and technoculture are combining to mobilize transformative alternatives to mainstream media, politics, economics, and formal education itself. While also used for hegemonic ends, as well as "technological terror," surveillance, and cyber-war (Kellner, 2003b), people have deployed new media technology—which encompasses the Internet, computers, cell phones, PDAs (personal digital assistants), digital cameras and recorders, and GPS (global positioning system) devices—to orchestrate the alter-globalization and antiwar movements, new political organizations and protests, along with novel oppositional forms of Situationist-inspired culture like flash mobs. Moreover, at times, emergent forms of online community utilizing blogs, wikis, social networking sites like Second Life, Twitter, Facebook, and MySpace, as well as public video forums such as YouTube, have attempted to further express a democratic social and educational project that involves the mass circulation of information and production of a worldwide knowledge culture on behalf of ecological politics.

Thus, the important role contemporary digital technology has had in developing contemporary pro-sustainability praxis must be underscored, as many Internet-oriented political and cultural projects today have an educational component through which they are reaffirming or reconfiguring what participatory and democratic forms of planetary citizenship will look like in the global/local future.

Paulo Freire: Promethean Pedagogy

> Dialogue cannot exist, however, in the absence of a profound love for the world and for people. The naming of the world, which is an act of creation and re-creation, is not possible if it is not infused with love. No matter where the oppressed are found, the act of love is commitment to their cause—the cause of liberation. And this commitment, because it is loving, is dialogical. As an act of bravery, love cannot be sentimental; as an act of freedom it must not serve as a pretext for manipulation. It must generate other acts of freedom; otherwise, it is not love.
> —Paulo Freire (2001)

While a plethora of work in English exists that looks to Paulo Freire's work for guidance on issues of literacy, radical democracy, and critical consciousness, there has arguably been less interest in the fourth major platform of the Freirian program—the problematization of technology as a modernization tool. Though significant divides clearly exist between rich and poor within the advanced capitalist nations of the north as well, this gap in the literature of critical pedagogy undoubtedly results from the differing political and economic needs of the southern countries in Latin America and Africa. Freire saw these development needs as a postcolonial problem—technology transfer could either be negotiated more on the people's own situational terms or those of corporate and state technocrats from the global north, who happily provide technical solutions in return for ongoing power and profit.

Akin to Freire's thinking, the present age of globalized technocapitalism and media spectacle increasingly requires a dialectical understanding of how new technologies are affecting the political economy in both overdeveloped and underdeveloped regions as part of a conjoined process. As Manuel Castells (1999) emphasizes, we need a critical theory that can "account for the structure of dependent societies and for the interactive effects between social structures asymmetrically located along the networks of the global economy" (p. 55). Therefore, as Peter McLaren (2000b) has noted:

> The globalization of capital, the move toward post-Fordist economic arrangements of flexible specialization, and the consolidation of neoliberal educational policies demand not only a vigorous and ongoing engagement with Freire's work, but also a reinvention of Freire in the context of current debates over information technologies and learning, global economic restructuring, and the effort to develop new modes of revolutionary struggle. (p. 15)

Notably, Freire himself echoed this sentiment in *Pedagogy of the Heart* (1997b), declaring that "Today's permanent and increasingly accelerated revolution of technology, the main bastion of capitalism against socialism, alters socioeconomic reality and requires a new comprehension of the facts upon which new political action must be founded" (p. 56).

A self-professed "man of television" and "man of radio" (Gadotti, 1994), Freire also believed in the "powerful role that electronically mediated culture plays in shaping identities, and the importance of the changing nature of the production of knowledge in the age of computer-based technologies" (Giroux, 2000, p. 153). Stating "It is not the media themselves which I criticize, but the way they are used" (Freire, 1972, p. 136), he should be considered a forerunner of the continually growing transdisciplinary field of critical media literacy. As early as *Pedagogy of the Oppressed*, Freire argued for the importance of teaching media literacy to empower individuals against manipulation and oppression, and using the most appropriate media to help teach the subject matter in question (Freire, 1972, pp. 114–16; 1998a, p. 123; Gadotti, 1994, p. 79). Hence, a re-examination of Freire's theory of education and technology is required in the context of the contemporary politics of mass and alternative media.

While Freire never developed a lengthy treatment of his views on computers and education, his work does contain a surprising degree of commentary related to the topic. Freire often employed cutting-edge media technologies as part of his system, even during his formative days as an educator in the early 1960s, and articulated his views on the politics of technology in a number of texts. Working in the tradition of Karl Marx, Freire propounded a dialectical view of technology (Freire, 1972, p. 157; 1997b, p. 35; 1998a, pp. 38, 92; Gadotti 1994, p. 78), in which he was always cautious of technology's potential to work as an apparatus of domination and oppression (Freire in Darder, 2002, p. xi; Gadotti, 1994, p. 79). Yet, he remained hopeful that it could also liberate people from the drudgery of existence, powerlessness, and inequality (Freire, 1993, p. 93; 1998a, p. 82). Thus, he notes in *Education for Critical Consciousness* (1973), "The answer does not lie in the rejection of the machine but in the humanization of man" (p. 35). In this way, Freire hoped to politicize the forces of science and technology (Freire, 1996), and thereby connect their popularization and democratization to a larger project of revolutionary humanism.[10]

Prior to the release of *Pedagogy of the Oppressed* in the United States, Paulo Freire was already famous in Latin America for being a radical educator whose innovative adult literacy programs made him first a Brazilian hero in 1962 and, soon thereafter, an enemy of the state who was jailed for a period and then exiled by military leaders after they took power via a coup d'état in 1964. His infamy resulted from his coordination of *cultural circles*, two-month-long literacy programs that were pronouncedly successful by combining training in reading and writing with lessons in self-reflection, cultural identity, and political agency. As director of the National Literacy Programme, Freire sought to deliver rapid literacy to millions of indigent people as part of a populist turn in Brazil's governing structure, which in turn threatened elite classes (and helped cement the coup) because Brazil's constitution then barred illiterate people from participating in the political process as voters. Freire's campaign, then, was an educational venture designed to transform peasants into citizens, significantly broadening the electoral base of the jobless, landless, and working poor, while empowering them to begin to speak and demand attention for their issues.

Importantly, the Freirian cultural circle made use of slide projectors, imported from Poland at $13/unit (Freire, 1973, p. 53),[11] which were used to display film slides that were the centerpiece of Freire's literacy training because of their ability to foster a collective learning environment and amplify reflective distancing (Sayers & Brown, 1993). For the slides, Freire enlisted the well-known artist Francisco Brenand to create "codified pictures" (Freire, 1973, p. 47) that were designed to help peasants semantically visualize the "culture making capacities of people and their communicative capacities" (Bee, 1981, p. 41). Composed of ten situations that intended to reveal how peasant life is cultural (and not natural) and thus human (and not animal),[12] Freire's film slides were displayed on the walls of peasants' homes, whereupon dialogues were conducted that analyzed the slides' various pictorial elements. The pictures themselves depicted a range of premodern and modern technologies, as well as other cultural artifacts, and the final slide ends on a meta-cognitive note by depicting a cultural circle session in progress.

Central to Freire's method was that once individual objects had been visually identified within the pictures, the words referring to them would themselves be projected in turn, then broken down syllabically, and finally, the phonemic families of the syllables would be revealed as "pieces" (Freire, 1973, p. 53) by which participants could construct new terms. In this way, after members of a cultural circle realized their ability to manipulate and create modern technologies through Brenand's pictures, they could transfer this knowledge to language itself and thereby recognize it as yet another technology available for their empowerment. Freire's intention, therefore, was to adopt technology pedagogically to demonstrate people's inherent productive and communicative abilities, as well as the possibility of their utilizing modern technologies critically and as part of a means to re-humanized ends.

Despite his early adoption of technology, Freire did not possess a naïve or technophilic attitude. To the contrary, in *Education: The Practice of Freedom* (1976) he is actually quite explicit about the tendency of high-technology and the electronic media to domesticate and maneuver people into behaving like mass-produced idolaters of technospectacle (p. 34). Under such conditions, Freire felt that:

> the rationality basic to science and technology disappears under the extraordinary effects of technology itself, and its place is taken by myth-making irrationalism....Technology thus ceases to be perceived by men as one of the greatest expressions of their creative power and becomes instead a species of new divinity to which they create a cult of worship. (Freire, 2000, pp. 62–63)

Reflecting upon this passage, Morrow and Torres (2002) correctly surmise that "Freire thus rejected from the outset any slavish imitation of given forms of 'modernization' driven by the unregulated capitalist exploitation of technologies" (p. 70).

In a less well-known text, but one deserving of being more widely read, Freire treats the theme of modernized development in a particularly rigorous manner as part of a sustained critique of neo-colonialism. Chronicling his activities in Chile during the late 1960s, the essay "Extension or Communication" sets out to address the question of whether the extension of modernized science and technology, exported to Chile (and other countries) as part of northern agricultural development initiatives, has served more to educate or alienate the traditionally based farming cultures of the Third World.[13] Though he was hardly unfriendly to Western modes of science and technology, Freire here inveighs against the politics of "cultural invasion" (Freire,

1973, p. 117), which in his mind amounts to the "imposition of one world view upon another" (Freire, 2001, p. 160). Cultural invasion, he notes,

> signifies that the ultimate seat of decision regarding the action of those who are invaded lies not with them but with the invaders. And when the power of decision is located outside rather than within the one who should decide, the latter has only the illusion of deciding. This is why there can be no socio-economic development in a dual, "reflex," invaded society. (Freire, 2000, p. 161)

Rejecting "the imposition of ostensibly value-neutral technocratic solutions on peasants that do not take into account either local knowledge or the impact on the community" (Morrow & Torres, 2002, p. 56), Freire defended the cultural integrity of *ethnoscience* and *ethnotechnology* (Freire, 1992, pp. 85, 227); but never in a *basist* (p. 84) manner.[14] Instead, he articulated a dialectical view in which the complex situation of autonomous Third World cultural practices, imperialist and capitalist First World desires, and the promise of modernity offered by the beneficial aspects of science and technology could be understood together as part of a holistic cultural development of radical *conscientizacao*. Often misrepresented as a "consciousness raising" project, in this context, the *conscientization process* (Roberts, 2000, pp. 144–45) is more properly revealed as a people's movement toward self-determination through engagement in emancipatory and critical praxis.[15]

Whereas agricultural and other technologies may have represented the leading-edge of a potential cultural invasion of the Third World in the 1960s, today similar debates rage around the attempt to develop a base of information and communication technologies (ICTs) throughout Latin America, Africa, and other regions of the planet. For example, the World Summit on the Information Society's 2003 *Plan of Action* targets that by 2015, with the help of the United Nations and the International Telecommunication Union, "all of the world's population will have access to television and radio services;" and that "half the world's inhabitants (will) have access to ICTs within their reach" (p. 2). In Freire's own work, the myriad possibilities and problems inherent in this vision were already beginning to be delineated and a critical politics was tentatively developed.

During the early 1990s, as Secretary of Education for the city of Sao Paulo, Freire recognized that computers represented a leading evolutionary line in the dominant society and thus he acted decisively to commit to the infusion of computers in all of the schools under his direction. As he told Moacir Gadotti:

> We need to overcome the underdevelopment Brazil faces in relation to the First World. We haven't come to the Department of Education to watch the death of schools and education, but to push them into the future. We are preparing the third millennium, which will demand a shorter distance between the knowledge of the rich and that of the poor. (Freire, 1993, p. 93)

Accordingly, Freire established the Central Laboratory for Educational Informatics while also investing in "televisions, video cassettes, sound machines, slide projectors, tape recorders, and 825 microcomputers" (p. 152). This is not to say that Freire sought to adopt computers uncritically; rather, his policy was formed as a result of a political and pedagogical strategy that sought to intervene in the status quo of a multi-mediated age. Though the rhetoric surrounding computers in education is often ebullient, Freire countered that he had worries about infused-technology, fearing "that the introduction of these more sophisticated means into the educational field will, once more, work in favor of those who have and against those who have not" (Gadotti, 1994, p. 79).

To this end, he was concerned that the science and technology of technocapitalism was increasingly producing knowledge representative only of "little groups of people, scientists" (Darder, 2002, p. ix). That most people, in either the First World or the Third, have neither the ability to produce a computer, nor even to manufacture or manipulate the software upon which computers run, was in his opinion antidemocratic and dangerously unparticipatory.

Hence, during a debate in the late 1980s with the computer-aficionado and educational futurist Seymour Papert, Freire rejected outright Papert's claim that computer technology surely meant the death of schools. Pointedly, Freire responded by observing that for all their pedagogical value and apparent historical necessity, computers were not technologically determined to compel students to use them in a critically conscious manner (Papert, 2000). Therefore, Freire felt that all cultures which now confront an ever-evolving and expanding global media culture have a responsibility to utilize new technologies with a critical (but hopeful) curiosity, thereby remaining committed to a pedagogy that both rigorously interrogates technology's more oppressive aspects and attempts through the conscientization of technology to foster reconstruction of the social, political, economic, cultural (or taken altogether—the ecological) problems that people face.

Ivan Illich: Epimethean Pedagogy

> We now need a name for those who value hope above expectations. We need a name for those who love people more than products....We need a name for those who love the earth on which each can meet the other....We need a name for those who collaborate with their Promethean brother in the lighting of the fire and the shaping of iron, but who do so to enhance their ability to tend and care and wait upon the other....I suggest that these hopeful brothers and sisters be called Epimethean men.
>
> —Ivan Illich (1970)

In contemplating Paulo Freire and Ivan Illich, Carlos Alberto Torres (2004) has written of the dialectical and complementary relationship between the two theorists, noting that the analogy that comes readily to mind is of Dr. Martin Luther King and Malcolm X. Equal in merit, but often opposite in approach, the work of Freire and Illich combines to provide a form of forward and backward looking Janus-figure. Both sought radically to defend the dignity inherent in humanity's potential and to provide the possibility of a better world and social justice, but the paths by which each pedagogue traveled largely diverged. Whereas Freire sought to intervene on behalf of the poor, critically pose problems into the "facticity" of their oppression, and divert technologies and other forms of cultural capital away from those in power toward those in need, the renegade priest/scholar/intellectual Ivan Illich developed a less messianic method. As an alternative to Freire's Promethean politics, Illich instead promoted an epimethean sentiment and style that looked to the historical past, and to the earth itself, for guidance in revealing the limits which, upon being transgressed, become counterproductive to life (Kahn, 2009).[16]

Though famous for his notorious *deschooling* thesis, which called for the dis-establishment of the social norm mandating institutionalized education, in later years Illich reconstructed his position by making it hostile to the idea of education in toto. Having previously realized that society's "hidden curriculum" (Illich, 1970) manufactures schools in order to introject forces of domination into student bodies, Illich went on to insist that, in a highly professionalized and commoditized media culture, all aspects of life either promote themselves as educative or increasingly demand some element of training as a cost of unchecked consumption. Under such conditions, the being possessing wisdom, *Homo sapiens*, becomes reduced to *Homo educandus*, the being in need of education (Illich, 1992a); and in an age when the computer becomes the "root metaphor" of existence (1992b), this reduction then becomes further processed and networked into the lost reality of *Homo programmandus* (Illich, 1995; Falbel in Hoinacki & Mitcham, 2002). Against this vision, Illich chose to defend "the fact that people have always known many things" (Cayley, 1992, p. 71) and have managed to live decently even amid conditions of hardship when left to their own autonomous devices. Thus, Illich came to propose a negative definition of *education* as the industrialized formula: "learning under the assumption of scarcity" (Illich, 1992a, p. 165).

One need not commit to Illich's indictment of education, however, to realize that one of his enduring contributions is the manner in which he perceived the deep ideological relationships between modern institutions like schooling, the church, factory production, medicine, the media, and transportation systems as authoritarian and dehumanizing elements of an unchecked industrial society. In a manner that seems quite congruent with Illich's thought, Marx (1990) wrote in *Capital*:

> In handicrafts and manufacture, the worker makes use of a tool; in the factory, the machine makes use of him. There the movements of the instrument of labor proceed from him, here it is the movements of the machine that he must follow. In manufacture the workers are the parts of a living mechanism. In the factory we have a lifeless mechanism which is independent of the workers, who are incorporated into it as its living appendages. (p. 548)

But for Marx, the alienation of the worker's productivity as it is subsumed within the industrial system through rationalized exploitation is not only inhumane but also an obstacle to the historical growth of human productive forces (Feenberg, 2002). Hence, in response, Marxist Prometheanism attempts to organize politically around normative demands for a more humane future that can only be realized, in part, through the liberated development of society's technical productivity. Illich's epimethean response to the inhumane industrial social system, by contrast, is closer to Audrey Lorde's (1990) declaration that "the master's tools will never demolish the master's house."

It is in this respect that Illich generally chose to speak of *tools*, and not technology or machines, both because it was a "simple word" (Cayley, 1992, p. 108) and because it was broad enough to:

> subsume into one category all rationally designed devices, be they artifacts or rules,

codes or operators, and…distinguish all these planned and engineered instrumentalities from other things such as food or implements, which in a given culture are not deemed to be subject to rationalization. (Illich, 1973, p. 22)

Therefore, for Illich, *tool* includes not only machines but any "means to an end which people plan and engineer" (Cayley, 1992, p. 109), such as industries and institutions.

In Illich's account, it is wrong to demonize tool making—he was practical, dialectical, and nontechnophobic—but tools do become problematical for Illich when they additionally produce "new possibilities and new expectations" that "impede the possibility of achieving the wanted end" (Tijmes in Hoinacki & Mitcham, 2002) for which they were made. Doing so, tools turn from being "means to ends" into the ends themselves, and they thus alter the social, natural, and psychological environments in which they arise (Illich, 1973). Remarking that "Highly capitalized tools require highly capitalized men" (p. 66), Illich implied that it is necessary that people struggle to master their tools, lest they be mastered by them. For when people uncritically operate tools that amplify human behavior and needs beyond the limits of the natural scales that existed prior to the tools' creation, tools move from being reasonably productive and rational to paradoxically counterproductive and irrational (Illich, 1982). For example, we see instances of this in the present development of the global communications network, in which members of society are subjected to the Moore's law version of "keeping up with the Joneses" to the extent that failing to remain technologically contemporary veritably excludes one from partaking of the dominant trend in social life generally. Ironically, for Illich this may be the hidden epimethean solution to the problem.

As Morrow and Torres (1995, p. 227) rightly observe, Illich's critique of counterproductive tools is thus related to Max Weber's concept of instrumental rationalization, as well as variant formulations proposed by Frankfurt School members like Max Horkheimer, Theodor Adorno and Herbert Marcuse.[17] For Weber, the process of instrumental rationalization resulted in the bureaucratization and disenchantment of existence, a sort of mechanized nullity brought about by "specialists without spirit" (Weber, 1958, p. 182). Likewise, Horkheimer and Adorno (2002) sought to critique the irrationalism produced by culture industries bent on reifying the rational in the form of fetishized commodities. Lastly, Marcuse (1964), in his notion of a "one-dimensional" world in which modern technology and capitalist instruments organize a society of domination in which any possible opposition becomes rationally foreclosed by it, posited the Frankenstein's monster of Promethean technologization in a manner quite comparable to Illich.

It is important to consider that those acting with epimethean values might respond quite differently than many contemporary social justice advocates to the problems outlined above. One avenue for political response would be to work to critically name the social system's various aspects and to march through its institutions, or to otherwise act transformatively at its margins, in such a way as to attempt to turn the potentials of the social mechanism toward the greater good. This "Dare to struggle, dare to win!" philosophy is quintessentially Promethean in character, however. For his part, Illich looked upon the growth of contemporary horrors like planned nuclear terror (Illich, 1992a, pp. 32–33) and the ubiquitous authoritarian reality of a dehumanized cybernetic "Techno-Moloch" (Illich, 1995, p. 237) as the necessarily catastrophic outcomes of a modern industrialism that has moved those who renounce it to a political position that is beyond words. As Adorno (2000) wrote, "To write poetry after Auschwitz is barbaric" (p. 210), and Illich similarly believed that the most moral response we might now make in the face of unprecedented socioecological crisis is to silently refuse to engage in debate about it and suffer it with grace and dignity.

Yet, Illich also remained married to hope for "postindustrial" conditions[18] and so he spent much of his life in imagining and creating *convivial tools* (Illich, 1973) that might reconstruct and transform the rampant technocracy and globalization of destructive industrialized culture that occurs under the moniker of modern development (Illich, 1971). Conversely, Illich's "tools for conviviality" are appropriate and congenial alternatives to tools of domination, as convivial tools promote learning, sociality, community, "autonomous and creative intercourse among persons, and the intercourse of persons with their environment" (Illich, 1973, p. 27). These tools work to produce a more democratic and sustainable society that is "simple in means and rich in ends" (Cayley, 1992, p. 17) and in which individuals can freely communicate, debate, and participate throughout all manner of a cultural and political life that respects the unique "balance among stability, change and tradition" (Illich, 1973, p. 82).

Through the idea of conviviality, Illich proposed positive norms to critique existing systems and construct sustainable options using values such as "survival, justice, and self-defined work" (p. 13). These criteria, he felt, could guide a reconstruction of education to serve the needs of varied communities, to promote democracy and social justice, and to redefine learning and work to promote creativity, community, and ecological balance between people and the planet. Indeed, Illich was one of the few critics working within radical pedagogy in his period who took seriously the warnings of the radical environmental movement and he critically appraised industrialized society within an ecological framework that envisaged postindustrial institutions of learning, democratization, and subsistence.

Illich was aware of how new tools like computers and other media technologies could themselves either enhance or distort life's balance depending upon how they are fit into a larger ecology of learning. He had a sense of computers' great promise, but was also suspicious of the new cybernetic regime of truth that seemed to him to be becoming instituted around ideas of data, networks, information, virtualization, feedback and transmission (Illich, 1992a). Thus, he remarked that he was fascinated by texts like Hofstadter's *Gödel, Escher, Bach* that epitomized the telemechanical aesthetic of artificial intelligence, but found them unreadable as they corresponded more to the "cut & paste" technics of word processing software than to a sequence of sentences representative of a continuous vision and inner-voice (Cayley, 1992). This underscored, perhaps, his chief fear of the information society: that mainstream computer literacy was outmoding the (as he saw it) eight centuries of print literacy that had given rise to moral subjectivity and the possibility of an individual's inner life (Illich, 1992a; 1992b). Illich saw as politically dangerous, and spiritually painful, that such interior texts were being exteriorized and broadcast upon digital screens.

On the other hand, Illich was "neither a romantic, nor a luddite" and he believed "the past was a foreign country" not worth endorsing (Cayley, 1992, p. 188). Nor did he believe that there was an either/or choice to be made between print and computer literacies; and so he suggested that for "anti-computer fundamentalists a trip through computerland, and some fun with controls, is a necessary ingredient for sanity in this age," as well as "a means of exorcism against the paralyzing spell the computer can cast" (Illich, 1992a, p. 207). Thus, Illich himself—ever the polymath—remained committed to learning and better understanding the latest developments in computing and while he personally chose to forgo word processing (as well as a regular relationship to newspapers, television, and automobiles), it is important to note that he was in advance of many intellectuals by making a great many of his books, essays, and lectures freely available for reading and sharing online.[19]

Further, while the last decade has produced a plethora of writing that cites Gilles Deleuze and Felix Guattari's concept of the "rhizome" as demonstrative of how the Internet can unlock radical possibilities in education, Illich's "learning webs" (Illich, 1971, pp. 72–104) and "tools for conviviality" (Illich, 1973) even better anticipate the Internet's various social networks, blogs, wikis, chat rooms, listservs, social networks and compendious archives in many respects. Thus, whereas big systems of computers promote modern bureaucracy and industry for Illich, personalized computers made accessible to the public for their own ends could demonstrate how online tools might provide resources, interactivity, and communities that could help revolutionize education by enhancing autonomous modes of learning. Consequently, Illich was aware of how technologies like computers could either advance or distort pedagogy depending on how they were fit into a well-balanced ecology of learning.

Reconstructing Education with Radical Pedagogies

> The only chance now lies in our taking this vocation as that of the friend. This is the way in which hope for a new society can spread. And the practice of it is not really through words but through little acts of foolish renunciation.
> —Ivan Illich, quoted in Cayley (2005)

Theorizing an ecologically democratic and multicultural reconstruction of education in the light of Freirian and Illichian critique demands that we develop theories of the multiple literacies needed to empower people in an era of expanding media, technology, and globalization. It appears certain that industrial, computer-based, and other digital technology will continue to drive educational and political trends in the decade to come. This means that if we choose not to abandon it altogether (and, following Illich, why not?), we should at least make sure that it works to enhance sustainability and the democratic empowerment of people, not just the corporate sector and privileged techno-elite who are generating ecological crises on a vast scale at present. Producing a sustainable citizenry and readying the conditions for the next generation's political struggle over how to respond to planetary ecocrisis should be a major goal of the reconstruction of education in the present and so represents a compelling form of ecoliteracy.

The development of convivial tools and a radically democratic ecopedagogy based on multiple technoliteracies must also aim to enable teachers and students to break with an entrenched paradigm of monomodel and homogenized instruction and instead engage in the reconstruction of education for a different kind of learning. Without seeking to foreclose what such pedagogy might look like, drawing upon Freire and Illich, tentative models can be envisioned. For instance, caring, dialogical, and transformative social relations in critical learning situations would promote civic cooperation, democracy, and positive cultural values, as well as fulfill human needs for communication, esteem, and being politically free with one another. But, an ecopedagogical educative reconstruction should also imagine its work as something other than "classwork"—in other words, it must help to foster a form of citizenship that is better able to negotiate the complexities of everyday life, labor, and culture, amid a dawning future that appears positively dangerous in so many ways. Such citizenship is at once powerfully local and rooted, tied to seemingly immediate conditions. However, the conditions of transnational capitalism also demand that we maintain a dialectical view of community that is international and planetary. We can no longer deny that our problems are *theirs*, and theirs are similarly *our own*.

As Freire (1998b; 1972) reminds us, critical pedagogy comprises the skills of both reading the word and reading the world. In problematizing new technologies and multiple literacies, then, we must constantly raise questions as educators, for example: Whose interests are emergent technologies and pedagogies based upon using them serving? Are they helping all social groups and individuals equally? Who is being excluded and why from decision making about the costs and benefits of the technology industry? Moreover, beyond seriously questioning the extent to which multiplying technologies and literacies serves simply to reproduce existing inequalities in the present, we must learn to strategize about the ways in which we might produce and reinhabit the lived realities of a more sustainable form of society amid the colonization of life by all manner of unsustainable technological paraphernalia. To my mind, these sorts of questions are eminently both Freirian and Illichian, regardless of the differences between their possible answers.

NOTES

1. See, respectively, http://upload.wikimedia.org/wikipedia/commons/2/25/Woody_Guthrie.jpg and http://blogs.amnesty.org.uk/uploads/blogs/entries/3127.jpg.
2. While I believe the common elements of this camp of work can be demonstrated to have been a major focus of Illich, by using this term I neither intend to reduce the outputs of these theorists to his ideas or to suggest that any or all of them would acquiesce to being

cast under this term.

3. For example, I believe this is true of some of the essays included in Bowers & Apffel-Marglin (2005).

4. Morrow & Torres's encyclopedic *Social Theory and Education* (1995) is notable for attempting a critical assessment of both Freire and Illich, though the context of the book's focus upon theories of cultural reproduction leaves a dialectical comparison of the two, and a close analysis of their thoughts on technology in relation to the recent growth of computing, beyond its scope.

5. As I note further in this chapter, Herbert Marcuse's theory of technology and politics undoubtedly exerted influence upon Illich, as it did for Freire, who in fact cites Marcuse in *Pedagogy of the Oppressed*. Students of Freire and Illich, then, should concern themselves with Marcuse's theories in order to better understand their generative aspects. For the contributions of Marcuse to education in particular, see Kellner, Lewis, Pierce & Cho (2008) and Kellner, Lewis & Pierce (2008).

6. The concept of *rational reconstruction* offered by the critical theorist Jurgen Habermas (1984) also deserves mention, but should not be conflated with the more experiential and dialectical project of reconstruction outlined here. More so, projects of Freirian (Freire, 1997a, p. 56; Morrow & Torres, 2002, p. 31; McLaren & da Silva, 1993, p. 69) and Illichian reconstruction (Illich, 1973) are obviously crucial, though my task here is to reconstruct them in terms of one another and contemporary needs in the context of present diverse situations in different locales.

7. For more on this, see Kahn (2005).

8. See http://nces.ed.gov/fastfacts/display.asp?id=46.

9. See www.apple.com/education/powerschool/profiles/lennox/.

10. Note the comparison to the discussion of a radicalized Enlightenment project of education, conceived as a critical *humanitas*, by Herbert Marcuse in his essay "The Individual in the Great Society" (2001, pp. 77-8).

11. Some years later, in Freire & Davis (1981), Freire placed the figure of each projector at $2.50. Considering the value of the dollar at that time, and that Freire purchased 35,000 units, this is obviously a large discrepancy in cost. Either way, one might surmise that Freire was comfortable with spending large sums of money on technology as long as it was being purchased for a progressive cause.

12. As I have written (Kahn, 2003), Freire's emphasis upon the dichotomy between human culture and animal nature must be understood as both an ideological tenet of Freire's radical humanism and as a reconstruction of the oppressive biases held by those in power that have historically labeled people of differing race, class, and/or gender as akin to "animals" in a "state of nature." Freire correctly perceived that when regimes dehumanize people and reduce them to uncultured savages it is part of a political project to deny them power. However, from a theoretical perspective Freire can be critiqued for mostly maintaining a

nondialectical view of humanity as being that which is not nonhuman and animal, which arguably served to underwrite a dichotomy between the politics of culture and nature in his thinking as well.

13. Freire's book is especially sophisticated because, though based in his practical attempts to deal with the real cultural and political problems besetting Chile at that time, Freire also speaks allegorically to the theoretical struggle between conservatives' attempt to delimit education as *educare* (the Latin root meaning "to cultivate" or "train" like a plant, an animal, or a child) and progressives' alternative vision of education as *educere* (meaning "to develop" that which is latent within). Thus, Freire undertakes an analysis of the modernization of agricultural practices wondering if the extension of modern science and technology into Chile should be better understood as a literal attempt to train the Third World in First World cultivation techniques (e.g., the way in which one trains a vine or disciplines a child), or as an attempt to help develop within the Third World its own latent abilities toward cultivating greater productivity and freedom in the face of modern science and technology.

14. Freire thus presciently theorized the critical, postcolonial research methodology of *cultural interaction* (Fay, 1996, p. 231), which serves as the ideological basis of some notable recent ethnoscience collections (Nader, 1996; Figueroa & Harding, 2003).

15. Freirian conscientization should thus be interpreted as a form of political engagement parallel to the de-colonial, but developmental and modernization-oriented "consciencism" formulated by the revolutionary African leader Kwame Nkrumah (1964, p. 70). As noted by Peter Roberts (2000, p. 138), Freire inherited the term *conscientizacao* from the archbishop of Recife and Olinda, Dom Helder Camara—whom Illich also studied under, resulting in his introduction to Paulo Freire (Cayley, 1992, p. 205).

16. Note the discussion of Prometheus and Epimetheus in the introduction, pp. 23–24.

17. However, their assertion that "Illich's whole theory is grounded in Marcuse's *One Dimensional Man*" (Morrow & Torres, 1995, p. 227) possibly obscures Illich's ability to synthesize a wide range of philosophies of technology, as well as his own novel contributions that made him a leader in the radical, alternative, and appropriate technology movements of the 1970s. Especially important influences upon Illich's theory of technology include Murray Bookchin, Jacques Ellul, Marshall McLuhan, Walter Ong, Leopold Kohr, E. F. Schumacher, Lewis Mumford, John McKnight, and the twelfth-century monk Hugh of St. Victor.

18. Such would be entirely different than found in the hyper-industrial society theorized as postindustrial by someone like Daniel Bell.

19. For instance, a large collection of Illich's writing is freely available at: http://ivan-illich.org.

Chapter Four

Organizational Transformation as Ecopedagogy: Traditional Ecological Knowledge as Real and New Science

> The message is: the privatization of knowledge, and of biodiversity, is a threat to the future of humanity. It's an enclosure of the intellectual and the biological commons, and we need to recover it. Simply because we need biodiversity and knowledge to continue to live.
>
> —Vandana Shiva (1997)

Introduction

The United Nation's Decade of Education for Sustainable Development (2005–15) calls for the integration of environmental science across the curriculum at all learning levels. Key questions must be raised, however, about the nature of environmental science as a discipline. At a practical level, for instance, "greening the academy" (Kahn & Nocella, Forthcoming) is becoming increasingly faddish throughout the United States but to what real transformative effect? Campuses everywhere are hiring sustainability managers, or lower-level technical administrators whose job is to document for college presidents and provosts how their campuses are fiscally responsible users of cutting-edge sustainable technologies, even when the truth is often something other still. Further, programs of environmental studies are being regularly developed throughout higher education; but while this field of study is supposedly interdisciplinary and interventionist, the reality is that it is almost always housed within natural science departments that usually interpret it (whether due to practical necessity in chasing grant funding or ideological biases formed through previous faculty training) as an opportunity to teach a curriculum of general environmental science with a small smattering of supportive ethics thrown in for good measure. Even the field's new nationwide professional society, the Association for Environmental Studies and Sciences, unnecessarily—in fact, as I am arguing here, redundantly—emphasizes the centrality of physical and biological science to environmental studies issues. Its name is not, for example, the Association for Environmental and Social Studies much less the Association for Environmental Studies and Critical Praxis.[1]

Of course, scientists have, can, and will certainly produce powerful analyses that contribute to the social good. The point I seek to make here is not antiscience. But mainstream institutional science can also be shown to be a form of culturally determined knowledge that often works with socially and environmentally deleterious assumptions about what science is, who gets to do it, and to what ends (Harding, 1998). In this respect, postcolonial and feminist critics of science have labeled the dominant paradigm of contemporary scientific research with the acronym WMS, which can mean both *Western modern science* and *white male science* (Pomeroy in Cobern & Loving, 2001). Thereby, WMS as an abstract, value-free set of universally falsifiable truths,[2] is recast as a specific sociocultural and political project that has—however unconsciously—securing the base of hegemonic power as one of its functions.[3] This includes (but is not limited to) naming WMS's complicity or co-construction with European-American processes of imperialism (Third World Network in Harding, 1993; Harding, 1998), racism (Gould in Harding, 1993), patriarchy (Cohn; Rose, Daniels in Wyer, et al., 2001), and the domination of nature (Perry in Wyer, et al., 2001; Scott in Nader, 1996).

Research scholars in environmental studies must therefore struggle to determine the field's methods and goals so that they are positioned dialectically in a critical liberatory relationship to WMS, if environmental studies is ever to be realized as an ecopedagogy capable of exhibiting what Sandra Harding (1991) has called "strong reflexivity" (p. 163) in its approach to knowledge production. Yet, too often sustainability is uncritically organized on campus such that it fundamentally accords with scientific types of technicism, instrumentalism, positivism, and naïve empiricism. As demonstrated by Hyslop-Margison and Naseem (2007), all of these approaches to science have become standards within a neoliberal educational paradigm. But, as just stated, the prevailing academic form of sustainability-as-environmental-science even more troublingly serves to root it institutionally to an unsustainable history of political and cultural domination that privileges "high status" (Bowers, 2001) forms of knowing (and knowers) associated with development of Eurocentrism. Of course, the two problems are hardly separate: we cannot divorce the enclosure of the commons from the desecration of peoples and the land (LaDuke, 2005; Bowers, 2006b). Thus *environmental science*, or *ecology*,[4] comes to be defined within a tradition of WMS through the active denial of the status of "real science" (Harding, 1998) to dominated peoples' knowledge claims and practices. This results in their continued cultural or political marginalization and oppression, as well as the exacerbation of ecological crises (Bowers, 2001, pp. 77–125; Snively & Corsiglia, 2001).

Such marginalization is blatantly exemplified by the academic organization of environmental studies vis-à-vis indigenous peoples' traditional ecological knowledge (TEK), or what Berkes (1993) defines as ways of being, wisdom, and cultural continuity "acquired over thousands of years of direct human contact with the environment" (p. 1). For sure, there are some developments within Western modern science that strike affinities with indigenous TEK perspectives (Deloria & Wildcat, 2001). It is additionally the case that there are attempts to develop culturally relevant forms of environmental science that are inclusive of and useful for tribal realities, such as the Ecological Society of America's SEEDS program, the Alaska Native Rural Systemic Initiative (Barnhardt & Kawagley, 2005), and the science education models provided by Gregory Cajete (1999a; 1993). Moreover, it is important to note that despite crucial differences between WMS and TEK, there is also an historical continuity between them.

Critical scholars of WMS have in fact pointed out how its evolution as a body of ideas and practices has depended to a large degree upon the assimilation (i.e., stealing) of knowledge from non-Western and indigenous traditions (Harding, 1993; Cajete, 2000). Where Western modern science differentiates itself then is through its ongoing tendency to either reject (or misunderstand) the cosmological and cultural conditions associated with TEK, along with the other knowledge systems that WMS surveils, as these knowledge systems become appropriated by WMS to serve the highly instrumental and technical ends of the global hegemony.[5] But despite WMS's involvement in the politics of dehumanization, a common humanity girds and serves to provide continuity between WMS and TEK as culturally different ways of knowing the world.

In other words, indigenous peoples must not be confused with their characterization by white society as "ecological Indians" (see Harkin & Lewis, 2007)—exoticized stereotypes of noble savages happily living in premodern ways and conditions within a state of nature. Indeed, the belittling of indigenous cultures through the West's representation of them as "savage" and "barbaric" (often through overt programs of education) is part of a colonial politics that has had odious historical consequences for indigenous peoples (Smith, 2002). Rather, tribes today are generally involved in the multifaceted mediation of revitalizing their long-standing cultural traditions with the contextual demands made by contemporary lifestyles and technologies as part of a progressive statement of their ongoing and collective strength (Lomawaima & McCarty, 2006). Therefore, while the continued manifestations of TEK are arguably the closest thing we have to truly sustainable varieties of knowledge—having developed through long-standing affiliations with nature that environmentalists themselves have argued should be the "bioregionalist" (Sale, 1985) outcomes promoted by ecological science—it is certainly true that not all indigenous peoples or cultures have engaged in ecologically sustainable ways of life at all times.[6]

For all of these reasons, TEK productively exhibits both useful similarities and discontinuities from WMS that should result in its easily being adopted as a primary element of environmental studies curricula everywhere. It might prove shocking therefore to find that indigenous students and faculty, as well as traditional ecological knowledge as a discourse, are highly underrepresented (i.e., absent) in environmental studies classes on many campuses. This is not a surprise because, as mentioned, the field of environmental studies tends to conform with or support Western modern science discourse in its theoretical outlook and curricular activity. Thus, when it is offered in higher education at all, TEK becomes the purview of Native American and American Indian studies programs or is otherwise included as parts of coursework in anthropology. In this way, both environmental studies and the larger academic disciplinary structure through which it is organized fail to meet the requirements of a multicultural and socially just approach (Sleeter & Grant, 2009) to education for sustainability.

Accordingly, ecopedagogy works strategically for TEK to be taught unabashedly as science in order to achieve a redistribution of "the cognitive and social benefits of scientific and technological changes" (Harding, 1998, p.

168) along more equitable and sustainable lines, while also reducing the sociocultural and environmental costs often brought on by the introduction of such changes. In so doing, ecopedagogy supports transformative research into who is excluded from the canons of sustainability scholarship, the methods it undertakes, and the normative sociopolitical frameworks of WMS generally. I emphasize that such research is *transformative*, it does not just want to generate answers to questions in these areas of interest, but also aspires to reflective action on and within the academy for a more just and ecologically sound society (see Grande, 2004; Malott, 2008; Deloria & Wildcat, 2001). That is, through scholarly dialogue with counter-hegemonic research standpoints on sustainability and democracy, ecopedagogy seeks out a type of science that allows for a reconfiguration of the geopolitical locations in which legitimate research takes place, who does it, and how.[7]

Traditional Ecological Knowledge: An Example

One such location, which I will next chronicle, is the Shundahai Network's Peace Camp: an antinuclear gathering of resistance, under the aegis of Western Shoshone leadership, which also protests the federal government's appropriation of tribal lands in order to create the Nevada Test Site (see Solnit, 2000). In recent years, I had the opportunity to visit the Peace Camp and conduct participatory action research with gathering members on matters relevant to the ecopedagogy movement's work. I find the camp startling because it offers a profound juxtaposition between its own subsistence-styled, sustainable TEK and the Armageddon-like WMS of the U.S. military, which has exploded nearly 1,000 atomic weapons at the Test Site since 1950 (including over one hundred above ground detonations that sent radioactive mushroom clouds high enough into the atmosphere so as to be visible from Las Vegas some ninety-five miles away). But the Peace Camp is meaningful also because it presents a picture of a contemporary TEK that is startlingly different from the New Age commodity spectacles of indigenous knowledge practices, what Aldred (2000) calls *plastic shamanism*, that increasingly populate white society and research. Thus, even though the camp centers on activities such as the Sunrise ceremony and traditional sweat lodge,[8] practices more typically associated with TEK, it also provides highly innovative models of planetary citizenship and of the oppositional uses of new media or cutting-edge renewable energy technologies that are not.

In the rest of this chapter I thus seek to argue two causes. First, that the

Peace Camp's TEK is a kind of real science that deserves to be upheld as legitimately teachable by fields like environmental studies and education. This does not mean the simple assimilation of indigenous knowledge practices into the dominant materialistic, reductionistic, and positivistic paradigm of modern Newtonian/Cartesian thinking proper. Rather, I envision this inclusion taking place as part of an institutional paradigm shift that radically reconstructs WMS for "the emergence of new perspectives, understandings, sensibilities, values, and paradigms that put in question the assumptions, methods, values, and interpretations of modern sciences" (Best & Kellner, 2001, p. 143). As Best and Kellner additionally note, this form of reconstructive project accords with Herbert Marcuse's theoretical call for a revolutionary politics that aims at the social production of a "new science" for "new sensibilities" (Marcuse, 1969; 1964). As I will argue in chapter 5 (see pp. 139–40), there is good reason to believe that Marcuse's new science should be interpreted as the knowledge and practice that is being developed out of the historical realization of what might be termed *a radical ecology of freedom*. It is my contention here, then, that the Peace Camp's TEK represents a compelling occurrence of such an ecology, and that it is therefore not just a real science but a type of new science that displays TEK situated in the present age as a living knowledge tradition engaged in and providing for sustainable cultural interactions.

The Peace Camp: TEK as Real Science

> We are one people. We cannot separate ourselves now. There are many good things to be done for our people and for the world. It is important to let things be good and it is important to teach the younger generation, so that things are not lost.
> —The reported last words of Western Shoshone spiritual leader and Shundahai Network founder, Corbin Harney (2007)

Since 1994, if one heads by car approximately an hour northwest from the Vegas strip and its consumptive mega-spectacle of simulated environments, gambling, and sex, one can arrive at a small, makeshift settlement in the arid desert on the outskirts of Death Valley. Just off the highway and directly across from the Camp Desert Rock entrance to the Nevada Nuclear Test Site run by the Department of Energy, Bechtel Corporation, and Wackenhut security forces, sits the Shundahai Network's Peace Camp. Created at the request of Corbin Harney, a spiritual elder and leader of the Western Shoshone tribe until his death in 2007, the camp is a place of convivial assembly where activists and organizations from all over the planet come to organize for another sort of world in the face of the many problems now threatening this one.

The social and cultural diversity to be found at the Peace Camp is profound. Quite literally one will find all manner of indigenous land rights, antinuclear, peace, environmentalist, antiwar, social justice, anarchist, and animal rights activists, along with university organizations, artists, and even spiritualist groups like the Raelians. All of these various communities of struggle come to honor the Western Shoshone's fight, learn from one another, and engage in opposition to the Test Site's broadly aggressive legacy. Hence, the events at the camp denote a critical pedagogy of border crossing where activists of many regions or nations dialogue strategically across their own identity-based cultural and geopolitical boundaries. Moreover, this solidarity building ultimately leads to a mass congregation at the Test Site's gated welcome station where, in order to reclaim the land as a sacred place for the Shoshone (and the earth), people literally cross into the zone controlled by the U.S. military-industrial complex—a 1,350 square mile area of illegal occupation. Though this has yet to drive the government off, such action fits nicely within the larger quilt of stories offered by Winona LaDuke (2005) of how today's indigenous communities are powerfully "recovering the sacred" by renaming the nature of their places. In this case, the guards, administrators, and nuclear scientists stationed behind the large "Keep out!" signs that adorn the barbed wire fencing of the Test Site are labeled as "trespassers" of Western Shoshone sovereignty and "profaners" of the spiritual responsibility to renew the soil.

Deploying numerous forms of highly participatory and performative dissent, with an emphasis upon direct action tactics, the Peace Camp compares favorably with other recent manifestations of organized antiglobalization resistance. Both similarly collect a wide-range of groups to protest global technocapitalist power in the further attempt to realize alternative modes of community. A major difference between them, however, is the principal and pervasive nature of indigenous TEK at the Peace Camp. Indeed, perhaps the single most important activity that happens there (according to Harney himself) is the building and maintaining of the sweat lodges that Shoshone leaders invite people to inhabit daily.

As Harney speaks of in his book, *The Way It Is* (1995), the lodges constitute a form of indigenous scientific research and medicine that is central to the Western Shoshone's understanding, practice of life, and relationship to nature as indigenous people.⁹ Far from being merely a low-tech sauna, the sweat lodges are a place of community healing, religion, politics, and education taken altogether—a truly sustainable "science for the people," so to speak.¹⁰ Indeed, when one contrasts the sweat lodges with the radioactive WMS that has toxified the area (spots around the Peace Camp continue to set off Geiger counters) and destroyed life throughout the region over the last half century, the lodges' practical and symbolic value as TEK becomes obvious.

One Science or Many?

The argument provided here most likely would fail to convince a nuclear physicist, or even a left-liberal environmental scientist within the academy, that sweat lodges and other forms of TEK are properly a form of legitimate science. This is for a number of ideological reasons. For starters, the dominant view of science relies upon the idea that "the material world ultimately judges the adequacy of our accounts of it" (Matthews, 1994). In this respect, then, WMS practitioners might ask how it can be materially demonstrated that sweat lodges cleanse indigenous (or any other) peoples of radioactive toxins, or otherwise serve to produce harmony and balance between those who sit in the sweat lodges with community members and the very forces of nature itself.

Some WMS practitioners might admit that the Peace Camp's lodges can work as sites of research within the context of the Shoshone's own belief system, but then they would argue that sweat lodges are little more than folk practices that may be meaningful anthropologically but which are unable to generate the sort of value-free, systemic knowledge from practical research that deserves the moniker of science. To the nuclear physicists (and those that accord with their definition of science as WMS), examples of TEK like medicinal sweat lodges are ultimately merely subjective and cultural in nature. As with churches, they should be considered places where groups can locally agree to believe in and define reality however they wish, but not such that their knowledge is legitimately thought transferable to the "real," predictive, controllable, and universally ordered world of objective science (see Stanley & Brickhouse, 2000).

If groups with subjective belief systems do make the transgressive claim to maintaining sciences through their beliefs, by WMS's standards they are consequently shown to be either sorts of fanatical true believers or others who are unable to meet the methodological rigors and necessities of "truly" scientific claims.[11] At absolute best, forms of TEK like the sweat lodge become cast by mainstream scientists as an "ethnoscience" (Cobern & Loving, 2001, p. 54), a kind of racially inflected "science, jr." or almost-science of the non-white world.

Yet that *science* should be reductively defined so as to be equivalent only to WMS is simply not born out by historical examination. Etymologically, *science* hails from the Latin word, *scientia*, which meant "knowledge in the broadest possible sense" (Snively & Corsiglia, 2001, p. 9), and the much narrower definition of *modern science*—meaning the WMS practiced and understood by experts indoctrinated into a cult of mathematical abstractions—is mainly a creation of the twentieth century. At most, the strong distinction between theoretical proof and merely practical knowledge that typifies aspects of WMS does not emerge until the beginnings of industrialism in the 1700s, and the connection of science to technological power over nature does not generally occur prior to 1850 (Nader, 1996).

To be sure, something like a Western practice of science has developed over time, dating back in fact to the ancient world, but there is no one clear origin or set of criteria that would allow us to trace the beliefs girding WMS either linearly or exactly back through history in anything resembling a meaningful fashion. Further, to reiterate a point made previously, all manner of TEK and related "alternative" scientific traditions (e.g., alchemy) have also informed and been appropriated by WMS whenever it was deemed profitable or otherwise advantageous to do so. We bury parts of this exploitative colonialist history when we fail to emphasize that behind the stark political and cultural differences of WMS and TEK, they also share a historical continuity that refutes WMS as an ideologically exceptionalist project.[12] Thus, like any ideology WMS has falsely naturalized itself and now uses its institutional and other forms of hegemonic advantage to reproduce further social, cultural, and environmental inequities (even when scientists from this tradition, sometimes quite sincerely, want to do work on solving exactly these problems).

Reconstructing Science with TEK

For the reasons outlined, the ecopedagogy movement must relate critically to science defined narrowly as "a naturalistic, material explanatory system used to account for natural phenomena that ideally must be objectively and empirically testable" (Cobern & Loving, 2001, p. 58). We should also illuminate how WMS has defined notions and projects of objectivity, rationality, and universalism toward its own interests over time (Figueroa & Harding, 2003). Instead of allowing researchers and scientists to perform, as Donna Haraway (1988) has put it, "the God trick"—or "the view from nowhere" where their voice hovers in a panoptic, apparently value-free quasi-omniscient space—ecopedagogy seeks to highlight that scientific knowledge is produced from political standpoints (see Harding, 2004). Standpoint methodologies demand that scientific research affirms its socially constructed nature, is done by people who belong to particular organizations, cultures, or nations, during particular historical moments, and is part of either a mainstream or marginalized political actuality (Hubbard in Wyer, et al., 2001; Fox Keller in Wyer, et al., 2001). Ecopedagogical researchers will thus want to think about the ways in which different cultures know and interact with nature's order generally, trying the best we can always to side with and begin from peoples' standpoints-from-below in terms of discursively exploring what (along with where, when, how, and why) it is these groups' know.

Dolores Calderon (2006) writes, "For educational research to be relevant for Native communities, I argue it needs to facilitate the continuation and development of indigenous epistemological or knowledge systems. In the United States, though, there is a historical legacy, continued in educational discourses and practices today that refuse to acknowledge indigenous knowledge systems" (p. 131). I believe that the ecopedagogy movement should actively work for the reconstruction of scientific literacy as part of the reorganization of education so that it is more fully equitable to all groups—human and nonhuman. We must especially demand transformative cognitive praxis in the academy on sustainability issues.

With this in mind, I conclude that teaching TEK (as takes place at the Peace Camp, for example) is a legitimate science in higher education. Further, it could become something of a science for the people capable of bolstering community in both indigenous and non-indigenous contexts.[13] This is not to say that hegemonic cultural and environmental terror will abate simply by granting institutional legitimacy to other forms of scientific research besides WMS, but it is my argument that until research scientists like high-level nuclear physicists really come to "know sweat," so to speak, that the removal of indigenous peoples from their land and the complete and utter desecration of the same will continue to be but little sweat off of the backs of the institutions these scientists presently serve.

TEK as a New Science?

> We have to construct the figure of a new David, the multitude as champion of asymmetrical combat, immaterial workers who become a new kind of combatants, cosmopolitan bricoleurs of resistance and cooperation. These are the ones who can throw the surplus of their knowledges and skills into the construction of a common struggle against imperial power....The democracy of the multitude needs a "new science."
>
> —Michael Hardt and Antonio Negri (2004)

Some of the most innovative political theory to have appeared in recent years is the work of Hardt & Negri (2004; 2000) on the globalized relationship between what they term *all-encompassing Empire* and the *multitude* that resists it from below and within. In short, they perceive Empire as a kind of new world order—an emergent sociopolitical network of forces, manifesting through the supranationalization of capitalism, that now moves us beyond the sovereign rule of the nation-state into horrors like the universalization of war, a powerful global police function, as well as authoritarian forms of control over all life and death. It represents a new series of ongoing exploitative acts that divide and conquer the land (as well as the communities that live there) on behalf of the spectacular growth of the infotainment society. In this way, it expropriates the planetary commons and privatizes nature in the name of a surplus value reaped primarily by the relative few. In service of these aims, it has increasingly developed highly technologized and instrumentalized forms of science. Indeed, they note science itself has become a pivotal scene of privatization, as alternative modes such as TEK and collaborative investigation are increasingly being subsumed under the singular ownership of corporate patents that is the dominant paradigm (Hardt & Negri, 2004, p. 151).

But the global flows of people, information, technologies, capital, energy, and so on that are the blood and veins of Empire are ultimately dependent upon, or represent, the productive capacity of the multitude that resists Empire in the common name of *cosmopolitan democracy*. In this sense, Hardt and Negri do not theorize Empire as simply the great fascistic evil from which there is no exit, but instead as a world state of capitalism that has emerged in response to the successful striving-from-below throughout the multitude's recent historical past. The multitude, therefore, stands in relation to Empire but cannot be fully subsumed under it. Indeed, it is actually Empire that is entirely dependent upon the productivity of the multitude for its existence, whereas the multitude is at least potentially an autonomous force (Hardt & Negri, 2004, p. 225). Thus, a parasitical image of Empire is formed—a vampiric Goliath hoping to dwarf its host (Hardt & Negri, 2000, p. 62). By contrast, a vision of the multitude is offered as a swarming, self-organizing posse that desires new communities, common names, aesthetic formations, and free relations based upon mutual understanding of internal differences (pp. 407–11; Hardt & Negri, 2004, pp. 91–93).

The challenge for the multitude, then, goes further than the demand that it resist Empire's technocapitalist machinery and the bureaucratic megamachinery of Empire's desire for a totally administered society. The multitude must additionally learn to reappropriate that which has been taken from it, to redeploy Empire's trappings toward its own ends, and to reconstruct a nonauthoritarian society that seeks the preservation of the common(s). The role of education and knowledge production is thus crucial for the multitude to realize a world of peace and freedom. Hence, Hardt and Negri call for practical, creative, and autonomous experimentation on behalf of a "new science" (Hardt & Negri, 2004, p. 353) of social being predicated upon the absolute nonreductive commonality of existential differences. Overtly, this is a challenge to and call for a reappropriation of the American Founders' "new political science" (Hardt & Negri, 2000, p. 161) of republican *paideia* (see chapter 1), as well as of today's dominant economistics—what Thomas Carlyle called the "dismal science" (Hardt & Negri, 2004, p. 157). But the new science of the multitude is also a conceit to Giambattista Vico's *Scienza Nuova* (1725), a foundational philosophy of history that claimed to surmise "the common nature of nations."[14] Besides being an influence on Hegelian/Marxist theory, Vico's philosophy is perhaps most famous for his conclusion that science must work to recover the greater truth that has become veiled in its myriad historical artifacts (i.e., the objects of the world). Doing so, he says, one comes to understand that "the true is the made" (*verum ipsum factum*): a constructivist principle that both declares that true knowledge of what is produced can only be held by the producer(s) themselves and that this is so because the act of production is an integral part of the process of knowing.

For this reason, Hardt and Negri have not gone so far as to offer specific models or blueprints by which to imagine the multitude, despite their lengthy theorization of it in numerous texts. Indeed, they have even gone so far as to stress, "Only the multitude through its practical experimentation will offer the models and determine when and how the possible becomes real" (Hardt & Negri, 2000, p. 411). Recognizing that by doing so I appear to commit myself to a violation of this principle, I believe it may prove useful however to close this chapter with a consideration of the Shundahai Peace Camp-as-multitude. Yet, I do not wish to suggest that the multitude can be reduced to any essential characteristics of the camp so much as to say that camp articulates a multiplicity of novel traits useful to understanding how the concept of the multitude can be productive for ecopedagogy as regards the organization of knowledge.

To this end, I want to argue that the Peace Camp is a self-organizing political power that realizes itself both through and in opposition to the forces of Empire. Even further, I believe it can be concluded that the camp—much like other forms of intentional ecovillage communities such as Paolo Soleri's Arcosanti, The Farm in Summertown, Tennessee, or the Findhorn Foundation in Morayshire, Scotland[15]—also represents a tentative attempt to live autonomously beyond the determinations of Empire for a dream of utopian sustainability. In these ways, by seeking to enact free relations of reciprocity with one another and the earth, camp members rename the commons and so become practitioners whose experiment in the Nevada nuclear desert is a kind of new science of ecological democracy for the twenty-first century.

The Peace Camp as Multitude

Using Empire

Although the Peace Camp directly opposes the brutal WMS of the Nevada Nuclear Test Site and offers alternative sciences like TEK, it is hardly premodern or even antiglobalization in its technical capacities. Again, the assemblage of groups and individuals—indigenous and nonindigenous both—that coalesce in any given camp experience includes people from a wide number of walks of life and from regions around the country and world (Kuletz, 1998). They know to arrive because of invitations circulated via the Internet, itself initially a military (and now largely corporate) technology; and despite the fact that the Shundahai website also offers people ways to organize ride sharing and resource pooling, most people still often must rely upon some combination of planes, buses, trains, and cars to travel to and from the camp proper. While it might be argued that trains or buses, for instance, are potentially consonant with future sustainability by offering mass transit, the pertinent issue here is that the transit system itself that has developed through and under the control of a large industrial corporate-state power structure is not. Rather it typifies the sort of administrative and technological decision making of modern business and government, even if these then largely moved to implement an even less sustainable consumer automobile paradigm post–World War II.

A second way in which the Peace Camp uses Empire relates to its on-site communications network. Once activists arrive at the camp, some members invariably set up short wave radio to coordinate and proliferate information about gathering events. Also, wireless Internet connections, along with solar and wind power to run these and other camp electronics, are established. The wind turbines could possibly represent something akin to technological self-subsistence, however the solar cells certainly do not, as photovoltaic technology is traceable to the silicon chip revolution of the last half-century. The machinery required to broadcast camp radio, as well as the computer servers and the like that are necessary to host the Internet there, are also products that can only be thought possible under modern industrial social conditions.

Finally, a third way that the Peace Camp reclaims elements of Empire for its own transformative agenda involves the work done at the camp by groups like Food Not Bombs and the Seeds of Peace Collective,[16] who provide for the camp's subsistence by serving free vegetarian and vegan breakfasts, lunches, and dinners. They are able to do this in large part because of their tactical ability to opt out of the "world food system" (Patel, 2008) through the collection of supermarket supplies that would otherwise be wasted in the name of new saleable goods, or by practicing freeganism[17] such as dumpster diving for acceptable rations that were already trashed. In return for mutual aid from those they feed, the groups' rescued foodstuffs are then redistributed back into the activist commons of the camp. Thus, they achieve a redefinition of the excessive squandering of groceries in a capitalist consumer society and make the negative externalities associated with the industrialization of food into structural forms of dietary welfare. But Food Not Bombs and Seeds of Peace Collective do this *within* the system, not wholly outside it. In this example, theirs is not a local permacultural food politics such as is now championed regularly by the environmental movement, but rather a re-localization of the transactional complexity behind transnational food for alternative ends.

Resisting Empire

The small-scale, self-contained Peace Camp is also intended to directly resist the full expanse of the Nevada Nuclear Test Site, its science and militarism, as well as the network of power that lies behind them. But the camp is not simply a formal protest in the usual sense. It is a defiant reappropriation of its place in the name of the sanctity of life, both natural and cultural.

Previously devastated by WMS, the land of the Peace Camp is actively cared for by camp attendees as well as blessed through TEK ceremonies. Whereas the federal government's previous requisitioning of Western Shoshone territory created a hard divide between private property outside the Test Site and public property (even if off limits to the public itself) inside, for a brief period of time each year the Peace Camp enacts a sort of "temporary autonomous zone" (Bey, 1985) of the commons as a third space of property relations. This resistance importantly occurs on both sides of the Test Site's barbed wire fence, demonstrating that the camp community intends to challenge both private and state-owned public varieties of real estate there.[18]

Compelling performative instances of cultural resistance also define the Peace Camp. As previously related, acts of TEK like the sweat lodge that take place during the gathering contain a clear oppositional impulse in this context. But this is true of the full panoply of creative interactions carried out by nonindigenous camp activists as well, examples of which include both formally scheduled and spontaneous song and dance (representing myriad styles and traditions), improvised theater and games, as well as workshops and other learning events. Indeed, the role of cultural elders and veteran activists teaching young children or other youth demonstrators particularly distinguishes the Peace Camp's ethos.

This intergenerational and pedagogical focus of the camp pointedly contradicts the culture of the Test Site, an adult-only facility where knowledge is shared only through formal channels and at the appropriate level of one's security clearance. Thus, the camp resists the desire to quarantine knowledge or otherwise lock it down by instead opening it up to a range of participatory and potentially democratic opportunities. Moreover, that camp situations involve a vital, autonomous aesthetic dimension also constitutes a form of refusal against the programmatic or top-down styled aesthetics favored by the government and military, which have principally fostered atomic annihilation at the Test Site over the last sixty years.

Toward a New Science

That the Peace Camp should emphasize modes of creative participation in its organization is hardly accidental. For, as Gregory Cajete (2000) teaches, creative participation is a primary element of native science generally. This serves to remind us that while the camp can (and should) be interpreted as a practical experiment in multidimensional reinhabitation—Hardt and Negri's multitude—here it manifests through the spirit of *shundahai*, the Western Shoshone term meaning "Peace and Harmony with all Creation." In other words, the Peace Camp community itself, as a self-determining, radically multicultural body attempting to realize ecological democracy, is an extension of TEK at its very foundations and across all manner of its various social activities.

As I have argued in this chapter, for the ecopedagogy movement this reveals two different avenues of transformative research into the organization of knowledge as the quest for a new science. First, by illuminating the way in which indigenous knowledge should be positioned chiefly at the forefront of supporting present and future generations of sustainability activism, an example such as the Peace Camp at the Nevada Nuclear Test Site challenges reified notions of TEK that one-dimensionally anchor its wisdom in the distant past. "Traditional" may connote senses of being old or time-tested, but TEK practitioners or supporters should work to explore the ways in which it more accurately denotes the ability to provide or deliver the conditions for ecoliteracy unto others in the present moment and the coming decades. That TEK has previously maintained this behavior across centuries at times, if not millennia, cannot be understated; however, it should not be allowed to occlude the ways in which TEK is alive and evolving in contemporary situations as a "Red pedagogy" (Grande, 2004). Learning to name, understand, and properly support the reimagination of TEK in this context is therefore a challenge that ecopedagogy must set for itself.

Another area for productive ecopedagogical research is to further explore the insight that TEK today takes on the form of the multitude in its resistance to maturing socioeconomic structures of unsustainable capitalism and global domination. Many advocates of TEK (especially its white romanticists) emphasize the role it can play in enlightening the sustainability movement with how to live on earth. It can; but as TEK informs this movement, it is itself also informed by it. TEK, then, must be thought as mediated by the wealth of cultural interactions increasingly taking place between a burgeoning transnational movement of indigenous peoples, on the one hand, as well as by indigenous peoples' encounters with the globalization of nonindigenous culture in all of its mainstream, subcultural, and alternative modes, on the other. Therefore, as the unfolding pedagogy of the multitude, TEK must be engaged as an ecoliteracy that is organized beyond mere identity politics. Yet, in the face of the long brutal history that is the dominant culture's appropriation of indigenous peoples' cultural well-being (LaDuke, 2005), ecopedagogy must extend great care not to reproduce or otherwise unthinkingly aid this process too, as it seeks through its research to illuminate the ways in which TEK, complexity theory, and a pedagogy for "total liberation" (Kahn & Humes, 2009) can increasingly be thought together.

Some of the major questions for such research revolve around apparently competing visions of democracy and sovereignty. In their call for a new science of the multitude, Hardt and Negri (2004) themselves oppose the creation of democracy to the destruction of sovereignty as an overarching demand. They write, "Sovereignty in all its forms inevitably poses power as the rule of the one and undermines the possibility of a full and absolute democracy. The project of democracy must today challenge all existing forms of sovereignty as a precondition for establishing democracy" (p. 353). We might wonder: Does indigenous sovereignty therefore differ ultimately from the sovereign rights claimed by George W. Bush, or now Obama (as well as many other world leaders both state and corporate), to invade, colonize, and destroy others' places at will because they have the authoritarian power to do so? Undoubtedly, Hardt and Negri would dismiss such a charge as baseless and deny that it is meaningful to liken U.S. imperialist sovereignty with the sovereignty maintained by First Nations. Yet, I am less certain that Hardt and Negri's theory to date can provide a viable defense of the preservation of indigenous sovereignty as an issue around which the multitude should coalesce in a sustained manner.

Correlatively, the critical indigenous scholar Sandy Grande (2004, p. 35) recognizes the potential value of radical democracy as an educative and political challenge to a transnational capitalist society, but she then distinguishes this democracy project from a sovereignty project of indigenization that she thinks can more appropriately serve native peoples. Indigenization is not to be confused with simply transplanting ideas of Western sovereignty into indigenous tribal realities. For, she further notes how some indigenous scholars like Taiaiake Alfred, a Mohawk political scientist, are usefully challenging commonplace Western notions of sovereignty in favor of the reassertion of prior indigenous forms (pp. 52–53). As I argue for a conception of TEK that is not just a legitimately real science but also a new science for ecological democracy in the twenty-first century and beyond, I similarly believe that revisioning concepts such as the commons and sovereignty from an indigenous standpoint is strategically crucial. Where I may differ from Grande's approach, however, is that I believe this can and should be done vis-à-vis the very idea of democracy as well.

As I have argued for it in this chapter, the TEK of the Peace Camp is neither a pure instance of indigenous sovereignty nor modern radical democracy. Instead, I see it as attempting to resolve the contradiction between these opposed polarities and so become both. This is to be achieved in part through the aspiration to detach and dethink the ideological history of Western democracy proper in favor of the reconstitution of democracy's indigenous form. As indigenous leader, Oren Lyons notes:

> In 1492, Haudenosaunee—which is better known as the Iroquois by the French, and Six Nations by the English—already had several hundred years of democracy, organized democracy. We had a constitution here based on peace, based on equity and justice, based on unity and health. This was ongoing.
>
> As far as I know, all the other Indian nations functioned more or less the same way. Their leadership was chosen by the people. Leaders were fundamentally servants to the people. And in our confederation, there was no place for an army. We didn't have a concept of a standing army, and we had no police. Nor was there a concept of jails, but there were of course fine perceptions of right and wrong, and rules and law. (in Lopez, 2007)

Thus, indigenous democracy entails and is entailed by indigenous sovereignty. The challenge now, then, is to understand and produce this relationship at the level of a multidimensional planetary emergence of which indigenous reality itself partakes and informs. In this way, TEK becomes a new science of sustainability—of the multitude as living, interactive, indigenous earth democracy—that rightly opposes all manifestations of Western sovereignty in favor of the reconstitution of an indigenization that authorizes and empowers all manner of creative, participatory, and peaceful planetary co-relations between beings great and small, human and nonhuman alike.

NOTES

1. This is not to say that there are not some within the association attempting to do critical

social work as environmental studies.

2. Modern science rests upon the methodological demands for falsifiability as set forth by Popper (1981; 1959).

3. This amounts to challenging what Figueroa & Harding (2003) refer to as the "unity-of-science" thesis of WMS. This thesis maintains "that there is one world, one 'truth' about it, and one and only one science that can, in principle, accurately represent that 'truth'" (p. 52). It also requires "that there be one and only one kind of ideal knower—presumably the 'rational man' of Enlightenment Liberal political philosophy" (p. 52). In modern times, these ideas about the nature and purpose of science have taken root so deeply that for a great many they are "a central component of their personal, professional, and public identities" (p. 54), a sort of morality that defines a specific type of community. This insight is important because it highlights the social aspects of such scientific belief, emphasizing its disciplinary and institutional character as "officially sanctioned knowledge which can be thought of as inquiry and investigation that Western governments and courts are prepared to support, acknowledge, and use" (Snively & Corsiglia, 2001, p. 9).

4. Sears (1964), Shepard & McKinley (1969), and Pena (1998) have argued that the science of ecology is inherently oppositional to modern development aims in its subversion of Western scientific reductionism through its emphasis on holism and biodiversity. While there are elements of truth to this claim, it avoids the underlying political economy and historicization of scientific ecology that has seen it used and promoted as a rational "wise use" profit-maximization strategy for industry by famous conservationists such as Gifford Pinchot at the turn of the twentieth century and as the engine of a form of technologically driven "green economy" that seeks to legitimate an entirely new era of investment in modern engineering and sustainable development now. Hence, it is more accurate to say that with other forms of science, ecological science contains within it the possibilities and limitations afforded by its historical moment, and that it is neither determinatively good nor bad, but rather depends upon the people who utilize and produce it to generate its relative ends. Lastly, it should be pointed out that despite potential problems with his definition as being nondialectical, Pena's (1998) definition of ecological science as the "traditional, place-bound, local knowledge of various unruly Others" (p. 3) mirrors the type of indigenous knowledge practices that I am attempting to legitimate in this chapter as sustainable and scientifically worthy.

5. Of course, it must be remembered that WMS is not a mere monolith and through its assimilation of various knowledge traditions, it has to some degree internalized struggle over what its ends should be, who it works for, who gets to do it and the like. Political splits within WMS were no more apparent than during the previous Bush tenure when the White House routinely silenced or edited state scientists' findings and recommendations when they ran afoul of neoliberal economic or neoconservative geopolitical interests.

6. Jared Diamond (2005), for instance, chronicles how Easter Islanders brought about their own social and environmental collapse as part of his larger message for the future of modern industrial society.

7. Here I believe ecopedagogical research must learn to draw from theoretical frameworks such

as Harding (2008); Lather (2007); Smith (2002); Sandoval (2000). The project I am advancing here fundamentally accords with the call for a critical science education curriculum advanced by Brickhouse & Kittleson (2006).

8. Despite the problematic rise in nontraditional, commercialized sweat lodge services designed to meet the needs of a wealthy and white marketplace (see Churchill, 1996, p. 228), I argue here for the cultural and political integrity of the Shoshone sweat lodge and other traditional Camp practices because they constitute critical learning experiences that I have had with Shoshone teachers under their direct leadership. Following Shirley Steinberg (in Malott, 2008, pp. 184–86), though, this does not absolve me of questions (buoyed by critical race theory) that make me wonder about the anthropological merit of my claims as a "white subject" able to partake of "traditional" knowledge. *These should and do cause me some concern* about what I believe I experienced as well as my ability to relate it here as useful to my theory, even if my experience was "authentic."

9. On the relationship between TEK and medicine, see also, Cajete (2000); Deloria & Wildcat (2001).

10. Lodges are run differently by different healers but can be found across nearly every form of indigenous community, where they possess a similar central cultural significance. Generally, branches or sticks are gathered to form a type of dome mound over a hollowed-out area in the ground or a naturally occurring chasm. River stones or other suitable rocks are gathered, blessed, and heated on a central fire until red hot. Sweats are generally single-sex, and men enter either naked or in shorts, while women are usually covered and refrain from partaking of sweat lodge ceremonies during their "moon" or menstruation period. Actual rituals involve blessings from the leader(s), chanting, and prayer—with the lodge symbolizing an earthen womb and an opportunity to make contact with the planetary spirit(s), which gave rise to and support humanity in its cosmic spiritual journey. Some sweats will pass a ceremonial pipe or other object in a circle among participants, allowing the holder to make an invocation or to otherwise address the lodge. Some sweat lodges permit the exiting of members who become overheated or otherwise panicked, though some do not. The intensity of the experience cannot be overstated and when one finally does leave the lodge, feelings of being reborn, cleansed, purified, augmented, strengthened, empowered, taught, healed, or otherwise liberated are extremely common. Indeed, even the therapeutic value of sweat lodges by certain aspects of Western society—such as drug/alcohol treatment centers and prisons—is well documented (Hall, 1985; Wilson, 2003; Waldram, 1993; Gossage, et al., 2003).

11. It should be noted that this is exactly the form of the scientific dismissal of the creationist-inspired pseudo-science of Intelligent Design. As Intelligent Design's political claim is to be considered *real science* per the terms of mainstream Western scientific standards, I agree with mainstream scientists that it can be refuted relatively easily as science in this way. Now, once one moves beyond the scope of demanding falsifiability as the determinative factor for judging scientific standards, as I am arguing for here with the Shoshone sweat lodge, it does occur to me that one potentially opens up a philosophical door in which all kinds of bogies such as I believe Intelligent Design to be might sneak through. Still, just because a knowledge practice need not be falsifiable in order to qualify as real science does not mean that *anything*

therefore can count as science—there are still a variety of real epistemological and sociocultural conditions that need to be evaluated of any given knowledge practice in order to determine its scientific value. Importantly, Pierce (2007) makes such an evaluation of Intelligent Design, as well as of mainstream Western science, and finds both to be insufficient representatives of a radically multicultural framework for producing scientific knowledge in the twenty-first century.

12. This exceptionalism says that "modern European American societies alone of all the world's cultures have managed to produce transcultural, universally valid claims" (Figueroa & Harding, 2003, p. 56).

13. Some theorists of science education believe that to characterize TEK as science is to allow it to be co-opted by the hegemonic status quo. In interrogating a real politic of epistemological pluralism, as has been argued for here, these theorists feel that indigenous knowledge and practice is best characterized within its own domain, as a form of alternative knowledge practice, so as to be able to critique the bias and exclusivity of WMS better from the margins. I am sensitive to this critique and we should be cautious of how, in using terms such as *science* or polemical labels such as *real science* to affirm TEK, a form of paternalistic and imperialistic control (even in the name of progressive democratic values) may be allowed to be once again re-inscribed over indigenous cultures (see Semali & Kincheloe, 1999). Further, following variants of postmodern theory, a coherent political goal could be to attempt to disempower singular paradigms of science in favor of the multiplication of alternative sciences on the whole. This chapter does not disagree with such theory, per se, but finds that the concrete politics of these matters is presently lagging behind such theoretical vision and so requires its own more limited strategic aims, however tentative.

14. This comes from the full title of Vico's work, *Principi di Scienza Nuova d'intorno alla Comune Natura delle Nazioni*.

15. See the websites located at http://www.arcosanti.org, http://www.thefarm.org, and http://www.findhorn.org, respectively.

16. See http://www.foodnotbombs.net and http://www.seeds-of-peace-collective.org/.

17. See http://en.wikipedia.org/wiki/Freeganism.

18. That it is the commons, in this case, does not mean that camp members fail to acknowledge it takes place within traditional Western Shoshone territory, however. Indeed, it is a category mistake to imagine the commons as an unbounded public domain in which everyone can stake equal rights to do as they wish (Hardison, 2006; Hardin, 1968). In this sense, the guardianship entailed by Western Shoshone sovereignty over the place allows for the commons to emerge.

Chapter Five

A Marcusian Ecopedagogy

> By saying no to the devastating empire of greed, whose center lies in North America, we are saying yes to another possible America....In saying no to a peace without dignity, we are saying yes to the sacred right of rebellion against injustice.
>
> —Eduardo Galeano, quoted in Espada (2000)

Introduction

In many respects, the twenty-first century has opened to the politics of the "no." The neoliberal and imperialist hegemons of the former Bush administration, their cousins once or twice removed now serving in Obama, Inc., as well as other key figures involved in expanding the U.S. market economy and military on behalf of the transnational class, have largely sought to erode or supersede norms of justice.[1] Thus, they have said "no" to legal protocols of war by abandoning the Geneva Convention, "no" to civil liberties and rights by rejecting the World Court internationally and domestically instituting (and then expanding in the face of widespread protest) the USA PATRIOT Act, and "no" to the regulation of capitalist greed by amending or repealing laws and other measures that were enacted to variously prevent corporate monopolism, renegade financial profiteering, industrial development beyond reason, and "natural resource" extraction beyond sustainability. Indeed, as this chapter will argue, the ruling class today promotes a ubiquitous sociocultural attitude that can best be described as the capitalist system's extinction of life generally in the form of a growing global ecological catastrophe.

In response, the populist grassroots have mobilized as decidedly antiglob-

antiglobalization[2] and antiwar, and their street slogans evince the negative character of the new social movements: "No blood for oil," "Not in our name," "No more years!" However, while the antiglobalization movement has incorporated Greens into its membership and been associated with important ecological battles such as Cochabamba, Bolivia's "water war" (Olivera, 2004), its aim has been more anticorporate than pro-ecology thus far. Likewise, though U.S.-led war has evoked ecological issues of crucial importance, such as the environmental effects of an oil economy and the widespread environmental toxicity produced through the American use of depleted uranium-enhanced weapons and vehicles, the antiwar movement (is there one left?) has largely evaded ecological critique in favor of anti-imperialist, antiracist, and pro-democracy discourses. The result has been an unfortunate failure to deeply integrate the environmental movement into contemporary progressive causes, and vice-versa, such that the socially educative potentials of what I have referred to as "a critical dialogue between social and eco-justice" (Kahn, 2003) have not materialized in the large.

Yet, such dialogues have begun to emerge in the radical margins of militant ecological politics, with affiliated organizations such as the Earth Liberation Front (ELF) and Animal Liberation Front (ALF) attempting to produce a revolutionary society based on critiques of the multiple fronts of systemic oppression (Rosebraugh, 2004; Pickering, 2002) as they move toward creating "interspecies alliance politics" (Best, 2003).[3] Having caused damage totaling more than $100 million over the last decade by most accounts, these groups have been labeled by the government as *ecoterrorists* and despite the truly violent activity of rightist white supremacist and antiabortion organizations are promoted as one of today's "most serious domestic terrorism threats" in the United States (Lewis, 2005).

While the charge of "terrorism" here is patently wrong and politically motivated (Best & Nocella, 2004, pp. 361–78), the government is correct that eco-militancy appears to be on the rise in the face of widespread environmental crisis and the utter failure of the mainstream environmental movement to offer successful opposition to the most rapacious aspects of capitalist development. Indeed, a 2005 RAND report even posits the greater convergence of the antiglobalization movement with ecological militancy over the next five years and predicts the potential "emergence of a new radical left-wing fringe across American society that is jointly directed against 'big business,' 'big money,' corporate power, and uncaring government" (Chalk, et al., 2005, p. 51).

It is in this context of movement resistance to our grave ecocrisis that Herbert Marcuse—the so-called "father of the New Left" and theorist of radical negation who emphasized the potential of the movements of the 1960s and 1970s to act as transformative educational catalysts in society—should be considered highly relevant.[4] Yet, Marcuse's philosophy seems mostly unnoticed by current ecological militants, as the movement is dominated on the one hand by the sort of pervading anti-intellectualism that Marcuse sought to educate among the New Left (Kellner, 2005a) and on the other by a linkage with questionable readings and uses of the philosophy of anarcho-primitivism.[5] Though groups like the ELF and ALF have been key in educating the public about the dangers and horrors of crucial ecological issues of the moment such as genetic engineering, urban sprawl, deforestation, automobile pollution and the effects of the oil economy, wildlife preservation, factory farms, and biomedical animal tests (Rosebraugh, 2004; Best & Nocella, 2004), they arguably lack a coherent theory of education and social revolution that could bolster and legitimate their advocacy.

This chapter, then, seeks to make (in however an introductory a fashion) a Marcusian intervention into the radical ecological politics of the present moment and thereby "educate the educators" (i.e., activists). As an explication of Marcuse's thought makes clear, groups like the ELF and ALF are undoubtedly social educators in that they hold key knowledge about the world that few possess and they have accordingly organized a politics (and to some degree a culture) that seeks to build upon and inform that knowledge. However, their politics run the risk of devolving into both a sort of vanguard elitism and despondent nihilism without a stronger theoretical basis, and Marcuse not only offers this but perhaps more than any other social theorist of recent memory combines the radical critique of society with a "positive utopianism" that can transcend pervading pessimism (Gur-Ze'ev, 1998).

But my attempt here to theorize a Marcusian ecopedagogy seeks to embody a sort of Marcusianism that moves beyond a straight explication that could run the risk of divorcing Marcuse's thought—itself always changing to meet the requirements of the present moment—from its sociohistorical context. In this way, Marcuse is hailed as an inspiration who is both a subject and object of the argument put forth here. Correspondingly, I will at times move beyond the conceptual language that Marcuse himself used in order to better intervene in present issues, all the while keeping the overall spirit of Marcuse's thought as a perpetual guide.

I begin by tracing the conjunction between the birth of radical ecological politics and the New Left, then move to a reconsideration of whether a Marcusian politics and culture of social intolerance is legitimate under contemporary circumstances. Following, I outline a call for the reconstruction of a Marcusian "pro-life" politics, and then close with a discussion of how Marcuse provides an under-utilized theory of politics *as* education and a revolutionary conception of *humanitas*, through which Marcuse sought to work to overcome the historical struggle and dichotomy between culture and nature, as well as the human and nonhuman animal. The conclusion offered is that Marcuse is a founding figure of a revolutionary ecopedagogy that says "No!" to the violent destruction of the earth, as it works to manifest a critical posthumanism based upon new life sensibilities that amounts to a utopian "Yes!" that will come to displace and end domination and repression broadly conceived.

The Modern Birth of Radical Ecological Politics

> I don't like to call it a disaster…I am amazed at the publicity for the loss of a few birds.
> —Fred L. Hartley, then-president of Union Oil Company, quoted in Clarke & Hemphill (2001)

In 1970, Earth Day largely marked the beginning of the modern environmental movement in the United States. Yet, a good case can be made that Earth Day itself, along with the sort of radical ecological politics now associated with groups like the Earth Liberation Front, erupted out of an event that took place the prior year (Corwin, 1989). While drilling for oil six miles off the coast of Santa Barbara on the afternoon of January 28, 1969, Union Oil Company's equipment failures resulted in a natural gas blowout from the new deep-sea hole they were excavating. Though the gas leak was quickly capped, the resulting pressure build up produced five additional breaks along a nearby underwater fault line (it is California after all), sending oil and gas billowing into the surrounding ocean.

Ultimately, it took the better part of twelve days to stop the main leaks, and some three million gallons of crude oil were released into an 800 square mile slick that contaminated the coastal waters, ruined 35 miles of shoreline, and damaged island ecologies. Amounting to a sort of Union Carbide disaster for nonhuman animals, over 10,000 birds, seals, dolphins, and other species were soon covered with tar, poisoned, or otherwise killed by chemical detergents used to break up the slick. Many more animals that did not die outright were adversely affected through destruction of their habitat, as the region became seriously polluted and took on the smell of the worst regulated oil refinery plant.

Santa Barbara's ecological catastrophe became a national media spectacle beamed into every American's television on the nightly news and, drawing on the nascent environmental consciousness sparked during the 1950s by Aldo Leopold's *A Sand County Almanac* and the 1960s by Rachel Carson's bestseller *Silent Spring*, public outrage erupted at the sort of governmental decision making that allowed Big Oil to cavalierly despoil the country for profit. It was revealed that oil companies had corrupted the U.S. Geological Survey, whose job it was to oversee the granting of offshore land leases and that such leases were routinely granted with little investigation as to their salience, save for that conducted by petroleum corporations themselves (whose data was private and could not be made a matter of public record). Further, corruption flowed from President Johnson's administration on down, as the Vietnam War was proving overly costly and so a policy of producing additional federal revenues from the selling off of natural resources was enacted in order to manufacture the illusion of budgetary economic soundness on the part of the country. As a result, the Santa Barbara channel had been auctioned off at the nice price of $602 million, providing the green light for oil companies to do with it as they willed, as the former proposal to turn the area into a wildlife sanctuary was quietly dropped from the agenda (Pacific Research Institute, 1999, p. 1).

Clearly, no one in power had ever stopped to question what the political effects of a giant slick in the Santa Barbara channel would be. A place of natural beauty that had been fighting as a community since the nineteenth century against the battleship-sized drill platforms stationed obtrusively on the horizon line, Santa Barbara was already mobilized on the issue. In the days following the spill, GOO (standing for Get Oil Out!) was created and it served as an organization to lead activist campaigns for reducing driving time, staging gas station boycotts, and burning oil company credit cards. Further, Santa Barbara was a city of wealth and intelligence. A home to many people with insider connections to alter the usual workings of the status quo, their pressure led to two major national policy changes: the enacting of a federal moratorium on leases for new offshore drilling (except in huge areas of the Gulf of Mexico) and the passage in 1970 of the National Environmental Policy Act (NEPA), the Magna Carta of environmental legislation in the United States. Finally, Santa Barbara was also a university town that was a hotbed of 1960s youth activism and counterculture.

The New Left-friendly community of Isla Vista, in particular, was known for its radicalism in opposing police repression, staging war resistance, and defending leftist University of California–Santa Barbara professors who were being denied tenure and removed from their posts (Gault-Williams, 1987). In 1970, Isla Vista militants responded with their own reply to the corporate energy cum military state by breaking into and razing the local branch of the Bank of America to the ground. The bank made a perfect target for many reasons. On the one hand, the bank was *the* community representative of capitalist business and, whether in its opposition to César Chávez's grape boycott or in its support for American imperialism (and hence the Vietnam War) through its opening of branches in Saigon and Bangkok, Bank of America was seen as corrosive to the community's social justice values.[6] But there is a less well-known, though equally important, reason that the bank was targeted. Bank of America directors were also known to sit on the board of Union Oil and so were themselves seen as responsible for the terrible oil spill of 1969 (Cleaver, 1970).

In this context, though the Earth Liberation Front's first American arson campaigns are dated only to 1997 (Rosebraugh, 2004), the torching of Isla Vista's Bank of America stands as one of the very first acts of uncompromising direct action to be found in U.S. environmentalism and thereby shows that radical ecological approaches to politics co-originated with the mainstream movement.[7] However, unlike the mainstream, Isla Vista New Left radicals tethered their ecological sensibility to an anticapitalist and antiimperialist stance that demanded a qualitative change in social relations. It was political moves such as this that served as an impetus for Marcuse in his end period to more straightforwardly announce the importance of ecological struggle as a central revolutionary theme.[8] Thus, groups like the ELF have a direct historical ally in Marcuse and so today's eco-radicals would benefit from a deeper investigation of Marcusian philosophy and its educational, political, and cultural implications.

Returning to the Question of Resisting Repressive Tolerance

Civil disobedience has many permutations.
You can block the streets in front of the United Nations.
You can lay down on the tracks, keep the nuke trains out of town,
Or you can pour gas on the condo and you can burn it down.
—"Song for the Earth Liberation Front," David Rovics (2004)

While there are dramatic differences between the political and cultural scene of the 1960s and the present, in many ways it seems like old times. Oil is again the center of political discussions. First, the Bush administration hunkered American forces down in a costly and apparently unwinnable Vietnam of its own making in Iraq, and now Obama is amplifying the war front in Afghanistan (with geopolitical access to increasing fossil fuel extraction in the region being at least one plausible reason as to why). Obama's Interior Secretary, Ken Salazar, recently set aside Bush's midnight ruling that would have opened up vast amounts of coastal waters to oil and gas drilling, but during the election season it was Obama's flip-flop on offshore drilling, in which he concluded that he no longer opposed it if it were done responsibly as part of a reformed energy platform, that helped Congress to end a drilling ban on federal waters that dated back to 1981. While a general moratorium on drilling remains in effect through 2012, oil and gas companies can presently begin exploring and studying a wide range of oceanic territory that was previously off limits. Meanwhile, Big Oil of course continues to work vigorously within the Beltway to gain full access to the continental shelf, among other potential exploration sites such as the Arctic National Wildlife Refuge (ANWR) in Alaska.

This is more than a little ominous, as the legitimacy of NEPA itself—the law created to make sure federal agencies properly account for potential environmental impacts prior to developing federal lands—was repeatedly undermined by the Bush administration, who sought to free industries from the law's time-consuming and expensive legislative regulatory procedure (Reiterman, 2005). Moreover, during this same time period, the nation bore witness to significant oil and gas disasters off our coasts. In 2005, a "mystery spill," expectedly unclaimed by any oil company, once again painted Santa Barbara beaches black and killed some 5,000 birds and other animals, making it one of the worst domestic oil catastrophes of recent memory (Covarrubias & Weiss, 2005). Less than a year later Hurricane Katrina's destruction of offshore refineries produced fifty large slicks along the Gulf Coast, rivaling the giant Exxon Valdez disaster in terms of oil spilled, as it became perhaps the greatest environmental catastrophe in the history of the United States.[9]

Yet, three and a half decades have also brought startling changes. Whereas 1969's spill both radicalized students into taking direct action against anti-ecological capitalism and galvanized a national environmental movement in the mainstream, 2005's oil slick passed by relatively unnoticed.

One might argue that in the present age, nothing short of the global warming mega-spectacle of movies such as the scientifically absurd *The Day After Tomorrow* or Al Gore's lecture-*cum*-political advertisement, *An Inconvenient Truth*, has enough emotional punch to break through the anaesthetized sensibilities of the seemingly oblivious masses.[10] In this sense, the relatively rare devastation wrought by a killer tsunami rouses widespread attention today, as the public passes by news about the toxic burdens brought to bear upon life by corporate and state malfeasance with little more than a bored shrug and, perhaps, a blog post.[11]

For sure, since 1999's Battle of Seattle the United States has seen a reinvention of public protest (Kahn & Kellner, 2007), and while people continue to link images of the sixties with notions of social discontent, the recent global antiwar protest of February 15, 2003 and the mass protests at the 2004 Republican convention in New York City (Kahn & Kellner, 2005) demonstrated dissent on a scale far beyond that ever mustered by the flower power youth. Still, why then did the counterculture of the 1960s seemingly accomplish so much while the contemporary left has appeared to suffer being overrun, consolidated, and ostensibly ignored despite its large numbers?

The answer requires a reconsideration of the past. Post–9/11 the United States has been engaged in a McCarthyesque crackdown on activists by brandishing them as terrorists, as corporations and the government intone treasured words like "freedom" and "democracy" (Best & Nocella, 2004). The state portrays itself as a security apparatus in charge of preserving the liberal ideal of tolerance, while it uses the extremism of groups like Al Qaeda to smear all of its enemies with charges of tyrannical fundamentalism. Thus, animal liberation activists like the SHAC7 are described as antidemocratic enemies of the state because of their willingness to directly challenge and attempt to shut down the self-imposed rights of corporations to cavalierly murder animals in the name of science and business, while Stop Huntingdon Animal Cruelty's opponents regularly promote themselves as good citizens who recognize the right to voice even the most unpopular opinions as long as those opinions do not step beyond the bounds of free speech into "intimidation" (Best & Kahn, 2005).

Herbert Marcuse wrote an important essay, "Repressive Tolerance" (1965), in which he examined the process by which the liberal state and its corporate members assert that they are fit models of democratic tolerance, as they insist that radical activists are subversive of the very ideals on which our society is based. In this essay, Marcuse notes that the claim that democratic tolerance requires activists to restrict their protests to legal street demonstrations and intra-governmental attempts to change policy is highly spurious. Tolerance, he says, arose as a political concept to protect the oppressed and minority viewpoints from being met with repressive violence from the ruling classes. However, when the call for tolerance is accordingly used by the ruling classes to protect themselves from interventions that seek to limit global violence and suppression, fear, and misery, it amounts to a perversion of tolerance that works to repress instead of liberate. Thusly, Marcuse thought such tolerance deserves to be met, without compromise, by acts of revolutionary resistance because capitalistic societies such as the United States manage to distort the very meanings of peace and truth by claiming that tolerance must be extended throughout society by the weak to the violence and falsity produced by the strong.

Many have criticized Marcuse for advocating violence against the system in order to quash the system's inherent violence (Kellner, 1984, p. 283). However, the critique of repressive tolerance is key to understanding why revolutionary violence would remain, if not ethical, a noncontradictory and legitimate mode of political challenge toward effecting "qualitative change" (Marcuse, 1968, p. 177).[12] For a tolerance that defends life must be committed to opposing the overwhelming violence wrought by the military, corporations, and the state as the manifestation of their power, and it is, by definition, to fail to work for their overthrow when one actively or passively tolerates them.

Marcuse therefore felt that revolutionary violence may in fact be necessary to move beyond political acts that either consciously or unconsciously side with, and thereby strengthen, the social agenda of the ruling classes. Further, he noted that the tremendous amount of concern (even among the left) evoked as to whether revolutionary violence is a just tactic fails to correlate to how often it is actually applied and practiced. Meanwhile, systemic violence constantly goes on everywhere either unnoticed and unchecked or celebrated outright. This goes to show, Marcuse felt, how hard it is to even think beyond the parameters set by repressive tolerance in a society such as our own, and this serves as yet another reason why such tolerance must, by any means necessary, be met with social intolerance.

Yet, Marcuse also recognized a wide-range of tactics, such as marching long term through the institutions,[13] grabbing positions of power wherever possible, and—in terms of ecological politics—"working *within* the capitalist framework" in order to stop "the physical pollution practiced by the system...here and now" (Marcuse, 1972a, p. 61) if they were undertaken with a revolutionary thrust toward a more sustainable, peaceful, and free planet.[14] On the other hand, Marcuse's key tactic has to be his concept of the "Great Refusal," which designated "a political practice of methodical disengagement from and refusal of the Establishment, aiming at a radical transvaluation of values" (Marcuse, 1968, p. 6). By rejecting death-principle culture and imagining an alternative reality principle based on reconciliatory life instincts capable of integrating humanity with its animal nature, Marcuse saw the Great Refusal from the first in ecological terms. This idea gripped the counterculture of the 1960s, who set out to create a plethora of new cultural forms and institutions (such as the environmental movement) across the whole spectrum of society.

Certainly, there are also bold new cultural forays in today's radical ecological politics. Increasingly, individuals and countercultural collectives are attempting to reject the mega-war-machine of the mainstream, as they take up veganism, permaculture, and other alternative lifestyles such as the Straight Edge movement that mixes urban punk stylings with a commitment to self-control, clean living, and political expressions like animal rights. Additionally, radical gathering events such as the Total Liberation Tour travel the country, and a variety of infoshops are actively investigating green political philosophies like social ecology and primitivism. Further, the last few years have seen a broad array of oppositional technopolitics (Kahn & Kellner, 2007; 2005). Blogs, wikis, tweets, txts, and websites are mushrooming everywhere to organize affinity groups, cover crucial issues dropped from the mainstream media, practice hactivism that jams corporate and state networks, gather otherwise secret information, and attempt to generate progressive (and sometimes even anticapitalist) culture.[15] Indeed, as to the latter, hardly an urban setting can be found that is free of some form of regular culture jam.

But as today's popular culture seems dominated by media spectacle and all manner of mass-commodified technological gadgetry as never before, eco-radicals must work harder still to distinguish the ways in which their culture represents a positive realization of anti-oppressive norms based on ideals of peace, beauty, and the subjectification of nature and is not just a nihilistic disapproval of a society that they may rightly deem unredeemable. That is, from a Marcusian perspective: A politics of burning down that lacks a correlative social, cultural, and educational reconstructive focus should not itself be tolerated.

Reimagining a Pro-Life Politics

> Be just and deal kindly with my people, for the dead are not powerless. Dead, did I say? There is no death, only a change of worlds.
> —Chief Seattle, quoted in Clark (1985)

George W. Bush was characterized as a pro-life leader for his desire to overturn *Roe v. Wade*, ban stem cell research, and stop funding for international aid organizations that offered counsel on abortions and provided contraceptives. Of course, in his role as outright war maker in Afghanistan and Iraq, indirect war maker through his administration's global neoliberal structural adjustment policies, and ecological war maker as the worst environmental president in U.S. history (Brechin & Freeman, 2004, p. 10), Bush is anything but pro-life. Rather, as the sort of über-representative of the affluent society, its forces, and its values, Bush was a fitting figurehead for the contemporary politics of mass extinction, global poverty, and ecological catastrophe. But, let us make no mistake about it, death-dealing politics such as Bush's extend far beyond the ideological confines of his neoliberal and neoconservative administration—it is at work on both sides of the aisle. So, from a perspective of radical ecology, mantras such as "Anybody but Bush" that liberals, left liberals, and other progressives attempted to use during the 2004 election cycle can be read as symptomatic of the need for wider education about the class-based and imperialistic nature of the dominant political and economic structure.

Marcuse himself referred to the sort of systemic disregard for life evinced by corporate states such as the United States as *ecocide* (Kellner, 2005a, p. 173)—the attempt to annihilate natural places by turning them into capitalist cultural spaces, a process that works hand in hand with the genocide and dehumanization of people as an expression of the market economy's perpetual expansion. More recently, others speak of ecocide as the destruction of the higher-order relations that govern ecosystems generally (Broswimmer, 2002), as when economies of need take areas characterized by complexity and diversity like the Amazonian rain forest and reduce them to the deforested and unstable monoculture of soybeans for cattle feed. However, while it is no doubt possible to disable an ecosystem from sustaining much life, it is not clear that one can actually kill it. Instead, we are witnessing a process by which bioregions are being transformed pathologically from natural ecologies of scale that support life to capitalist ecologies that function beyond limit and threaten death. In this way, the current globalization of capitalism that institutes classist, racist, sexist, and speciesist oppression is a sort of biocidal agent.

It is biocidal, also, in a more philosophical sense. The term *bios* is a Greek word that has come to designate natural life as studied by the science of biology. Originally, though, *bios* meant a sort of characterized life (Kerenyi, 1976, p. xxxii)—as in a "biography"—that is demonstrated by the active subjectivity of sentient beings. In this manner, organizations like People for the Ethical Treatment of Animals (PETA) have as their ultimate goal the social recognition of animals' *bios* (Guillermo, 2005) and, accordingly, want them to be afforded the status of being considered subjects of a life that are therefore deserving of rights. When compared with the larger socio-political context against which PETA struggles, however, the McDonaldization of the planet is obviously moving in the opposite direction. Most beings today, including the great earth and the sustaining cosmos beyond, are instead increasingly reduced to one-dimensional objects for exploitation, and should they provide too much resistance to the schemes of profit and power in the process, they are tagged for systematic removal.

In stark contrast to the objectification of life that typifies mainstream culture in the United States, as well as to the sense of life as "characterized" that is represented by the idea of *bios*, the Greeks (in a manner similar to many indigenous cultures) held that life was fundamentally *zoë*—a multidimensional and multiplicitous realm of indestructible being (Kerenyi, 1976). Thus, in Greek culture primeval and natural places were consecrated to the pagan deity Pan (whose name means "all"), and these were held to be sacred groves where *zoë* was especially concentrated in its power. The final point, then, is that ruling class politics are also zoöcidal, though not in the sense that it kills *zoë* (which cannot be killed by definition). Rather, in instituting a transnational network of murderous profanity over the sacred, in paving paradise in order to put up a parking lot, capitalist life is zoöcidal in that it seeks to colonize any and all spaces in which cultures based on understandings and reverence for *zoë* can thrive.[16]

The call, therefore, to those seeking to take up ecopedagogy is unmistakable. They must, if they are not doing so already, integrate the ecological critique into the politics and culture of civic freedom and equality and so become sustainability radicals.[17] Further, ecopedagogues themselves must increasingly move to develop cultural relationships to nature that exhibit the sort of positive liberatory values that have emerged out of a long history of social struggle and that Marcuse felt could be accessed through the subordination of "destructive energy to erotic energy" (Marcuse, 1992, p. 36) in the present age. Of course, ecopedagogues will also have to learn, grow, and of course teach, the values and practices that unfold a new sensibility toward life that emerges from the attempt to liberate and reconcile with the earth proper. In this respect, perhaps, the reimagination of a pro-life politics in which human and nonhuman beings are understood as both *bios* and *zoë* represents for us the great anticapitalist challenge of the current historical moment. In the face of expanding zoöcide, to think that this could occur without widespread rebellion and, ultimately, revolution, seems extremely doubtful. As Marcuse (1966) remarked: "In defense of life: the phrase has explosive meaning in the affluent society" (p. 20). Today, radical sustainability politics such as practiced by the ELF seem determined to prove Marcuse right.

Ecopedagogy as Political Education and Educational Politics

> The real change which would free men and things, remains the task of political action.
>
> —Herbert Marcuse (1972b)

To my mind, Marcuse is one of the preeminent philosophers of education in modern times, not only because he lived as well as propounded a radical theory of education as a centerpiece of his social critique and political plan of action, but because his educational theory was essentially linked to the ecological problem of human and nonhuman relations due to his understanding that education is a cultural activity, and that in Western history such culture has systematically defined itself against nature in both a hierarchically dominating and repressive manner (see chapter 1). As a result, Marcuse conceived education in both an intra- and extra-institutional scope, and ultimately saw it as incorporating all of social life and the total existential development of humanity toward the achievement of new life sensibilities and consciousness capable of "dispelling the false and mutilated consciousness of the people so that they themselves experience their condition, and its abolition, as vital need, and apprehend the ways and means of their liberation" (Marcuse, 1972a, p. 28). For Marcuse, then, education and revolution were largely synonymous forces, which struggled against their reified forms as one-dimensionalizing political apparatuses, corrupting professions, and dehumanizing cultural forms.

Recently, in a number of books and essays, Peter McLaren has become a leading voice in the call for and development of a "revolutionary critical pedagogy" that can heretically challenge market-logic and reformist ideology in favor of whole-scale social transformation. In fact, in an essay written with Donna Houston (McLaren & Houston, 2005), McLaren has even charted a sort of "eco-socialist pedagogy" that stands in defense of convicted ELF activists such as Jeffrey Luers, as it militates against what he terms the *Hummer* educational machinations of the mainstream and capitalist status-quo. However, where Marcusian erotic archetypes could deeply inform and bolster such pedagogy, McLaren has instead pointed to the symbolic (and other) influence of Che Guevara and Paulo Freire (McLaren, 2000a) and, most recently, to purveyors of the Bolivarian revolution such as Hugo Chavez (see McLaren & Jaramillo, 2007). Indeed, while it is worth reiterating that Freire himself finally recognized the importance of ecological struggle at the end of his life, writing that "It must be present in any educational practice of a radical, critical, and liberating nature" (Freire, 2004, p. 47), it can be argued that the U.S. educational left's reliance upon Freire over the last thirty years slowed pedagogical developments vis-à-vis the liberation of nature and nonhuman animals that Marcuse himself had posited as necessary for humanization by the 1950s and 1960s.

Ilan Gur-Ze'ev (2002) has pointed out how Marcuse promoted a form of German *Bildung*, or the cultural learning and practices that intend the shaping and formation of more fully realized human beings (Kellner, 2003c), as counter-education. Marcuse himself was more prone to speak of the goal of "humanity" and the ideal of *humanitas* (Kellner, 2001) or even the universal sense of human dignity connoted by *Menschlichkeit* (Marcuse, 1977), but always in a manner akin to *Bildung*. Hence, Marcuse extols an ideal of human potential and freedom that can emerge only through political action as education. As we have seen in chapter 1, historically educational projects like *humanitas* and *Bildung*, while serving emancipatory purposes also promoted self-contradictions of class privilege and other forms of oppression, yet Marcuse hardly utilized these conceptions in an idealistic manner and instead sought to use them as critical challenges to the educational and political status quo of the current day. Marcuse also enlisted them as utopian thrusts to explore and expand the Marxist conception of "human needs"—the full development of which is necessary for the appropriation of nature that would afford the realization of humanity as a *species being*—as being something more than an epiphenomenon of coming socialist institutions by rooting them in the universally instinctual (i.e., natural) needs of individuals (Marcuse, 1972a; Kellner, 2001). In this, species being itself ultimately opens up to other species in a common heritage and Marcuse's revolutionary humanism came to take the form of a critical posthumanism that advanced the hope for an end to anthropocentric oppression and exploitation of the nonhuman (Marcuse, 1972a).

Against those like Blanke (1996) who find evidence of a mystical consciousness in Marcuse's attempt to reconcile human culture with nature by liberating the latter as a subject in its own right, the correlative of the new sensibilities afforded by a qualitative change in society, Marcuse's thinking is nothing of the sort.[18] As with Horkeimer and Adorno, Marcuse recognized that the fundamental problem of society was the "Domination of man through the domination of nature" (Marcuse, 1972a, p. 62)—that nature was the primordial object whose subjection distinguished and founded human control. Thus, he concluded that the "realization of nature through the realization of man as 'species being'" (Kellner, 2001, p. 132) must logically represent the historical end goal of the movement toward liberation.

His point is, first, that education must seek to forge a new nature, which must be envisioned and aesthetically materialized because such would be the dialectical condition for the emergence of socialism and a new culture of human relations. Secondly, beyond what he sees as base Marxist accounts that leave even this form of nature as but a sphere of productive force for non-class-based social relations, Marcuse posits an ecology of freedom[19] that finds that as people start to live freely for their own sake and generate instinctual autonomy, this must be mirrored externally by the increasing relation to all that surrounds them in the spirit and form of freedom. Dialectically speaking, the liberation of the external environment and the production of peace and freedom also entail the potential realization of the subjective conditions that could be the basis of a "new science" capable of manifesting a free society.

It seems very apparent to me that if Marcuse were alive today, he would not hail New Age transcendentalism as a solution to the gross transnationalization of capital, the external human plight of over three billion, and the internal psychical plight of billions more still. Based on his ecological writings, I am led to conclude that he would be deeply alarmed by the unprecedented mass extinction of species, the waylaying of planetary ecosystems, and the mass production of zoöcide at levels that can soon no longer even profit the ruling classes, as they too are threatened.[20] Finally, I would like to imagine that Marcuse would have built on his ecological philosophy and politics to become a tireless promoter and organizer of a sort of ecopedagogy that is not a simple addendum to standard curricula, but rather an attempt to raze education under capitalism in favor of a pedagogy of the repressed that seeks to wage revolutionary political struggle toward a future culture based on radical notions of sustainability and a humanized nature that can represent values of tolerance, beauty, subjectivity, and freedom on a cosmic scale.

With the scale of suffering so nearly unimaginable and the politics of counterrevolution so fully in effect at the present, Marcuse might well highlight the marginal political and cultural actors, such as the Earth Liberation Front, who work to educate society as to the gravity of the consequences of their political economy and provide the hope of alternative relationships in and with the world. Without a doubt, in turning earth warriors into leading pedagogues (who, though, as this chapter has declared, nevertheless stand in need of their own education as educators), the Marcusian spirit has moved far afield from most contemporary educational discourse, even in ecological and environmental education. However, this may well be not because of the naïveté or insufficiency of the educational projects and political goals mounted by the earth or animal liberation movements, but rather because present versions of academic ecoliteracy are themselves seriously, and perhaps gravely, depoliticized.

NOTES

1. It goes without saying that I am not equating the crimes of Bush and Cheney with the fledgling Obama administration. On the other hand, the Obama administration *has* staffed itself with many neoliberal ideologues, has continued illegal war abroad (even exacerbating it in Afghanistan and Pakistan), has failed to prosecute the former administration for torture, has backed out on environmental pledges, and proved altogether moderate on many key issues demanding immediate action of a more radical nature.

2. Many people speak instead of the *alter-globalization movement* in order to highlight that the movement is not simply negative in its outlook. However, *antiglobalization* remains the most popular moniker and its negative character is arguably its most central feature to date. See Kahn & Kellner (2006).

3. The Earth Liberation Front "is an international underground organization that uses direct action in the form of economic sabotage to stop the exploitation and destruction of the natural environment" (Pickering, 2002, p. 58).

The Earth Liberation Front's guidelines are:

1. To cause as much economic damage as possible to a given entity that is profiting off the destruction of the natural environment and life for selfish greed and profit

2. To educate the public on the atrocities committed against the environment and life

3. To take all necessary precautions against harming life

4. Besides those texts highlighted earlier (see p. 47), other volumes exploring Marcuse's theory of ecological politics include: Blanke (1996); Gottlieb (1994); Merchant (1994); Luke (1994); and Kellner (1992). DeLuca (2002) has also written how the Frankfurt School offers a theoretical base for radical environmentalism but erringly overlooks Marcuse's work in favor of an analysis of Horkheimer and Adorno, as well as of Marcuse's student William Leiss.

5. On anarcho-primitivism see Jensen (2006, 2007); Perlman (1983); Zerzan (2002); and journals like *Green Anarchy* and *Fifth Estate*. I should note here that in many respects I think that no better critique of industrial society, from an ecological perspective, has been made than by these chief theorists who comprise the anarcho-primitivist movement. Even the Unabomber Manifesto revealed trenchant insights into the manifest problems in contemporary technological society, despite other theoretical failures contained therein. However, despite making the strongest statement of ethical rage over the contemporary destruction of the planet by "civilized," capitalist society, I feel that they fail to deal adequately and honestly with the existential situation faced by opponents of that society who stand in dialectical, if not directly substantive, relationship to the same. Hence, I believe Marcuse's theory provides a more fertile and consistent ground from which to produce ecopedagogical resistance, as it is thoroughly dialectical and reconstructive without being altogether accommodating or merely reformist of the prevailing ecological/technological/organizational social norms.

6. On the Isla Vista incident, see the 1970 documentary film, *Don't Bank on Amerika*, by Peter Biskind, Stephen Hornick and John C. Manning (Cinecong Films).

7. Others (Chalk, et al., 2005, p. 47; Jarboe, 2002) date the ELF as originating earlier in the 1990s, as an outcropping of Earth First!, the environmental group that counseled *monkeywrenching* as "resistance to the destruction of natural diversity and wilderness" (Foreman, in Foreman & Haywood, 2002, p. 9). However, monkeywrenching was specifically defined as *not revolutionary*, in that such acts "do not aim to overthrow any social, political, or economic system" (p. 10). Likewise, while the FBI connects the ELF to the birth of the Sea Shepherd Conservation Society in 1977, under the rubric of "special interest extremism" (Jarboe, 2002), Sea Shepherd's mission to conserve and protect the oceans, and its commitment to international law and the UN World Charter for Nature, disclose it as a non-revolutionary group different in kind than the ELF.

8. As proof of Marcuse's support of militant environmentalism beyond the mainstream, one should note the beginning to Marcuse's 1972 talk "Ecology and Revolution" (Kellner, 2005a) – a piece essentially dating, as we have seen, to the beginning of the U.S. environmental movement. In that talk, Marcuse begins by declaring, "Coming from the United States, I am a little uneasy discussing the ecological movement, which has already by and large been co-

opted [there]" (p. 173). In the context of the title referencing "revolution," Marcuse can only be deploring that American environmentalism was proving in its infancy to be a largely white and bourgeois politics that had as its goal governmental regulations that would afford some measure of humane reform while leaving the system basically unchallenged. Of course, Marcuse was not against meliorating policies that arose out of a revolutionary struggle, but his later point was that these should be considered one means toward a larger end, and not an end in themselves.

9. The Exxon Valdez oil spill occurred on March 24, 1989, when an Exxon-owned oil tanker struck a reef in Prince William Sound, spilling tens of million of gallons of crude oil. It is estimated that the deaths of birds, seals, whales, otters, and fish ran to the hundreds of thousands at a minimum as a result of this accident.

10. Marcuse called for a revolutionary aesthetic sensibility because he felt that capitalist culture served to anæsthetize people to the history of real needs (Reitz, 2000). Building upon Reitz, it can be suggested that media spectacles are required to generate feeling and enthusiasm in advanced capitalist nations like the United States much in the same way that substance abusers require larger and larger doses of pharmaceuticals in order to unlock the "high" that they crave. In other words, the addict's senses are reduced to low-levels of affectation as part of a process of ever diminishing returns.

11. This is not to say that blogging cannot be an effective and interesting form of technopolitics, even as regards ecological concerns (see Kahn & Kellner, 2008).

12. The concept of *qualitative change* is crucial in this respect, as Marcuse recognized that many political revolutions have sustained the "continuum of repression" and simply "replaced one system of domination by another" (Marcuse, 1968, p. 177). The revolution for qualitative change, however, has as its means and end the elimination of systemic violence in its myriad forms and the augmentation of beauty and happiness in the name of liberty and justice.

13. The "long march through the institutions" originated with the Italian Marxist Antonio Gramsci, but Marcuse integrated this concept/strategy by way of the radical Rudi Dutschke (who went on to help establish the Green Party in Germany). For Marcuse, this did not mean merely engaging in parliamentary democratic governmental processes, but it also required staging organized demonstrations for clearly identified issues, creating radical caucuses and counter-institutions, and—most importantly of all—in moving into the institutions of society, becoming educated in the work to be done, and educating others so that everyone will be prepared to manage these positions in a non-oppressive manner should the revolutionary moment arise on the world's stage.

14. Readers of Marcuse will no doubt know that in the early to mid-1970s he strategically modified his revolutionary position from the mid-1960s in order to deal with the apparent fracturing and staggering repression of radical groups that had begun to occur. Previously, he had uncompromisingly attacked repressive tolerance and called for examinations of how third and first world revolts might ignite a revolutionary subject(s) capable of overthrowing the capitalist status quo (Marcuse, 1968), but Marcuse's end period publications and talks often saw him advising that liberal society would have to be utilized from within (Marcuse,

1972a) through a sort of double-agency of insider/outsider status. In a lecture of this period entitled "The Radical Movement," for instance, he notes that "we are in a very bad situation" that means "there is a lesser evil" in which "even certain compromises with liberals are on the agenda" (The audio of this lecture is available online at: http://www.gseis.ucla.edu/faculty/kellner/ media/marcuse2.ram). Yet, Marcuse never abandoned his belief that violence against capitalist aggression was legitimate under the right conditions, and while he did not fetishize revolutionary violence, he did believe that in certain situations (at least) such as the advance of the revolutionary movements during the 1960s that such violence could be justified (Marcuse, 1968; 1965).

15. During June 2009, the latest example of such materialized in Iran where state authorities attempted to close down media outlets after initial domestic outrage occurred in the face of an apparently rigged presidential election. However, protesters used Facebook, Twitter, and cell phones to get video images out to the world of internal resistance and of the heavy police crack-down, which resulted in many deaths.

16. Note, that by employing a concept of *zoë* I do not seek to romanticize ancient Greece's ecological well-being. On the contrary, it is often cited that Plato himself appears to speak in the *Republic* of the environmental devastation wrought by the clear-cutting of the Athenian forests—one of the earlier historical accounts of such behavior on the part of people. Thus, my point is not that the Greeks were ecologically sound and the current age is not, but rather that Greek society, in spite of its environmental destruction, also developed rituals and practices (as well as articulated philosophies) related to a rich ecological sense of being as *zoë*, and that such could be resurrected today in opposition to the one-dimensionalized life of alienated toil and purposeless over-consumption and production. For a book-length treatment of this idea, see Lewis & Kahn (Forthcoming). In this respect, *zoë* should be likened to what Deloria & Wildcat (2001) denote as *power*.

17. For recent examples of this form of alliance politics, see Best & Nocella (2006; 2004).

18. See also Steve Vogel's "Marcuse and the 'New Science'" in Abromeit & Cobb (2004, pp. 240–46) for an attack on Marcuse's thinking in this respect that I am arguing here is based on a fundamental mis-reading of Marcuse's theoretical and political project. For additional critiques of Marcuse's ideas of *new science* and *new sensibilities*, including those infamously made by Jurgen Habermas himself, see Kellner (1984).

19. The phrase *ecology of freedom* was famously coined by the founder of the social ecology movement, Murray Bookchin (1982). Bookchin was undoubtedly an influence on Marcuse's critical ecological theory of the 1960s and 1970s, just as Marcuse and the Frankfurt School were an important influence on much of Bookchin's work. But whereas the Frankfurt School theorists dialectically related the domination of humanity to the domination of nature, and Marcuse spoke of the need for the liberation of both, Bookchin's position evolved away from domination of nature concerns and he instead posited that natural destruction can be solved only through the achievement of non-oppressive social relations between people. While Marcuse, I believe, would agree that such social relations are necessary preconditions for real peace, he also gestured to the agency of nature itself and in this manner more deeply

anticipated the radical ecological political vision of the present moment with its connection to animal and earth liberation ideas and values. Thus, I suggest Marcuse presents an alternative version of the theory of the ecology of freedom, albeit in snippets of books and essays only.

20. More than being alarmed, there is his own textual evidence to support the notion that he would condone the growth of vegetarianism and veganism as sociopolitical movements. Marcuse was a great lover of animals, with a particular fondness for the hippopotamus. While it is true that he declared the "campaign for universal vegetarianism" to be "premature" in the context of so much human suffering (Marcuse, 1972a, p. 68)—a sort of ranking of oppressions on his part—Marcuse felt that any society would seek to reduce animal suffering in direct proportion to its production of freedom generally. Today, when the political reality of animal suffering is so extreme, even defenders of more liberal views of animal welfare have moved to vegetarian and vegan lifestyles to protest the cruel realities inherent in practices such as factory farming. Further, that recent ecological studies have revealed that a move to a global vegetarian diet would also be key in reducing the suffering of human hunger, narrowing the economic inequities between nations, and lessening dangerous planetary phenomena like global warming would have been interpreted by Marcuse, I believe, as meaning that meat-based diets should be increasingly relegated to the past and that universal vegetarianism has begun to come of age.

Epilogue

A Concluding Parable: Judi Bari as Ecopedagogue

> Starting from the very reasonable, but unfortunately revolutionary concept that social practices which threaten the continuation of life on Earth must be changed, we need a theory of revolutionary ecology that will encompass social and biological issues, class struggle, and a recognition of the role of global corporate capitalism in the oppression of peoples and the destruction of nature.
> —Judi Bari (1997)

> Yes, they lined Joe Hill up against the wall,
> Blindfold over his eyes.
> It's the life of a rebel that he chose to live;
> It's the death of a rebel that he died.
> —Phil Ochs (1968)

While it has been the purpose of this book to begin to offer the philosophical foundations for a type of ecopedagogy that might be developed out of the dialectical critical theories espoused by figures such of Paulo Freire, Ivan Illich, and Herbert Marcuse as they are utilized to survey the world's present ecological state of emergency and the unsustainable state of contemporary mainstream educational practices, I would like to close specifically with a paean to the revolutionary Judi Bari as a kind of worded end-piece for the work. For the life and spirit of Judi Bari, perhaps more than any other recent figure I can think of—with the possible exception of César Chávez—*refleshes* (see McLaren, 1991, p. 162) the constellation of thoughts and feelings that I will argue constellate today's counter-hegemonic resistance movement for ecopedagogy.

Among other things, Judi Bari was a radical educator, a loving mother, an important grassroots theorist, a rebellious fiddler and song writer, a black belt in karate, a skilled carpenter and graphic artist, a seasoned antiracist, antiwar and anti-imperialist activist, a Wobbly organizer for the Industrial Workers of the World, and the woman who ushered in a new era for Earth First! as the principal leader of Northern Californian radical environmentalism from 1988 until her death in 1997. Just as funny, irreverent, and impassioned as the monkeywrenching patriarchs of Earth First! who came before her, like Edward Abbey (2000), Bari brought an additional ecofeminist sensibility to the movement that opened the door for women to become more involved in radical environmentalism and to take on leading roles in strategizing and protesting. For instance, in this respect the emergence of Julia Butterfly Hill in the late 1990s, herself famed for unifying diverse groups around her ecofeminist message and courageous two-year treesit in the giant sequoia named Luna, is only thinkable as part of the historical legacy created by Judi Bari.

Bari also promoted a wide-ranging ethics of care and maintained the central importance of respecting one's community, and these attitudes generally allowed groups associated with Bari to transcend the energy-draining, chest-thumping wars over theory and strategy that marred many other male-dominated environmentalist camps in the 1980s and 1990s (e.g., see Bookchin & Foreman's (1991) protracted, bitter, and unproductive argument). Further, Bari's syncretistic and dialogical approach to political life was capable even of healing enflamed divisions on the ground between environmentalists and logging company workers, and in this respect she single-handedly forged a united front between them in an allied struggle against the exploitative corporate establishment that appeared to care neither for old growth redwoods nor the people charged with felling them.

She also understood the need to draw upon a range of interest groups, even those without a clear self-identified stake in her campaigns, as part of a strong and diversified transformative movement. As Marcuse had done in the 1960s, Bari realized the key role that college students could play in fomenting revolutionary social change, and she spent countless hours driving to campuses in order to speak with student groups and encourage their political involvement. As a result, even in the remote regions of northern Mendocino and Humboldt counties, Bari's protests regularly drew thousands of sympathizers, and she was thereby revealed to be an unparalleled American agitator for radical democracy and ecological well-being heading into the new millennium.

For her hard work, in 1990, Bari was car-bombed when a device hidden beneath her seat exploded as she travelled through Oakland on an organizational tour for her annual Redwood Summer campaign. The force of the explosion shattered her pelvis and her body was riddled with shrapnel such that she lived out the rest of her days in severe pain. Amazingly, Bari herself immediately became the target of an FBI disinformation campaign in the press, where she was infamously described as a dangerous terrorist who had clumsily blown herself up with a device of her own manufacture, and the agency then joined forces with the Oakland police to arrest her accordingly.

This occurred despite Bari having filed numerous police reports in the weeks previous to the blast about anonymous death threats made against her. After the fact, when friends (and Bari herself) openly questioned why no investigation was forthcoming of the captains of the timber industry, whose power and profit stood to be directly undermined by Bari's actions, police only further slandered her and openly lied by claiming that parts used in the car bomb were an exact match to materials taken from a search of Bari's home.

Though Bari could hardly walk in the months after the attack, upon discharge from the hospital she quickly took to the streets and began the last of her great public battles, singing a new song—"The FBI Stole my Fiddle (and I want my Fiddle Back!)"—as she filed a federal law suit against the FBI and Oakland police in order to clear her name and reveal the truth about a conspiracy to silence her conducted by law enforcement agencies. In 2002, a jury finally exonerated Bari, charging the FBI and the Oakland police some $4.4 million in damages for false arrest, unlawful search and seizure, and the violation of her First Amendment right to freedom of speech.

This was, of course, a posthumous victory. For Bari, like Rachel Carson before her, died of breast cancer—an epidemic disease affecting approximately one out of every eight women in the United States. Feminist environmentalist Joni Seager (2003) advocates that breast cancer is a social and environmental attack on the health of women's bodies by the patriarchal establishment represented by the powerful array of institutions comprising the chemical and medical industries. Interestingly, akin to radical ecological educator Ivan Illich, who forewent professional medical treatment for his own ultimately lethal face tumor due to ideological reasons, Bari too chose to skip a therapeutic protocol defined by waves of chemotherapy and surgery.

Instead, Bari opted for an anarchic and honest death, one that could be lived freely with her loved ones on her own terms in opposition to becoming objectified by bureaucratic regimes of social normalcy bent on iatrogenic damage and control. Rather than lie in a hospital bed connected to tubes and morphine drips, she worked tirelessly to the end on her lawsuit, giving videotaped testimony for the record and organizing her legal files for future use. Bari understood very clearly that her historical role did not end with her physical demise—there is also an ecology of spiritual action in this world.

According to many friends, Bari remained true to her Wobbly background and the epitaph of Joe Hill with her final words: "Don't mourn. Organize!" This, then, is the spirit behind this book and the challenge for ecopedagogy generally as we move into the twenty-first century confronted by the crisis of unprecedented global ecological catastrophe. Now is not the time for our tears—especially as our silent spring grows ever closer to twelve months in length. We as educators must do all in our power to confront the powerful forces that appear ever increasingly to associate the drive for economic and political profit-taking with the all-out exploitation and annihilation of life on earth. We must not be silenced by the intensity of the threat. Rather, we must teach as if our lives depend on it, because in a very real way, they do.

In the book's opening, I explained how the ecopedagogy movement began in the global south and is now to be reimagined in a northern (and planetary) context around socially reconstructive issues of worldview creation, technology production, and the organization of knowledge. In closing, and so having come full circle, I would like then to give the final words to another voice of freedom and peace from the global south, Subcomandante Insurgente Marcos. While speaking on behalf of another movement that refuses the clear-cutting of hope, the Zapatistas, Marcos (2001) similarly articulates a beautiful vision for ecopedagogy in the years to come:

> To plant the tree of tomorrow, that is what we want. We know that in these frenetic times of "realistic" politics, of fallen banners, of polls substituting for democracy, of neoliberal criminals who call for crusades against what they are hiding and what feeds them, of chameleon-like metamorphoses, saying we want to plant the tree of tomorrow sounds foolish and crazy; but nevertheless, to us it is not a phrase born of drama or obsolete utopianism.
>
> We know all that, and nevertheless, that is what we want. And that is what we are doing....The tree of tomorrow is a space where everyone is, where the other knows and respects the other others, and where the false light loses the last battle. If

you press me to be precise, I would tell you it is a place with democracy, liberty, and justice: that is the tree of tomorrow. (p. 282)

Afterword

Mediating Critical Pedagogy and Critical Theory: Richard Kahn's Ecopedagogy

Richard Kahn's groundbreaking work *Critical Pedagogy, Ecoliteracy, and Planetary Crisis: The Ecopedagogy Movement* is distinguished by its merging of perspectives from a Frankfurt School–inspired critical theory of society with a broad range of figures within critical pedagogy to provide conceptual foundations on behalf of a critical ecopedagogy movement that is mounting both within the academy and the larger activist community. Kahn opens by evoking the magnitude of the current ecological crisis and a corresponding crisis in environmental education. He cites figures indicating that although most serious educators are in favor in principle of offering environmental education, the actually existing programs are few and are usually marginalized. Moreover, as Kahn argues, the dominant models of environmental education abstract the ecosphere from developments in the economy, science, and technology, and are generally uncritical of the existing society. Hence, they are unable to provide real insight into the causes of our ecological crisis and to mobilize on behalf of adequate responses.

In addition, existing eco-education all too often lacks solid philosophical and ethical vision, needed to discern the dialectical relationships between nature and culture as well as to produce forms of consciousness that recognize the importance of a sustainable society that is inclusive of all forms of life. Kahn argues that part of the ecological crisis is the historical development of an anthropocentric worldview grounded in a sense that nature is a stuff of domination to be used by humans to meet their needs and purposes. Hence, a critical ecopedagogy needs to be rooted in a critique of the domination of nature, of the global technocapitalist infrastructure that puts profit and market forces before humans, nature, and social goods, and of an unfettered Big Science and Technology that has instrumental and mechanistic perspectives on nature and that fails to see the need for a robust ecological science and appropriate technologies.

Kahn interprets ecopedagogy as a development of critical pedagogy that first took place in Latin America after the UN Earth Summit of 1992 held in Rio de Janeiro, Brazil which inspired educators like Paulo Freire to see that a critical pedagogy must cultivate an ecological dimension. Kahn points out that the later Freire (and his Latin American associates) perceived ecopedagogy as part of a planetary movement for social and educational change through work at the grassroots level in social movements and through global educational institutions such as the Earth Charter Initiative. Kahn recognizes previous efforts that began to articulate a critical ecopedagogy in North America and lays out the contributions of previous advocates of critical pedagogy and ecoliteracy, but argues for a broader philosophical and critical vision rooted in the works of Ivan Illich and Herbert Marcuse, as well as Freire.

To correct existing forms of environmental education, Kahn calls for a critical ecopedagogy that is concerned with understanding how political economy and ideology produce the domination of nature. A critical ecopedagogy promotes a dynamic and complex definition of ecoliteracy that seeks to promote the idea that while we are hemmed in by the limits of and interpolated by destructive institutional forms, we can recognize and transcend these thresholds through measures of individual transformation and collective action, which aim for sustainable place-based relationships. Fleshing out his project, Kahn engages an emergent tripartite model of ecoliteracy that involves interlocking forms of functional/technical literacy (e.g., environmental science), cultural literacy (e.g., which cultural practices/traditions further sustainability or hinder it?), and critical intersectional literacy focusing on the oppressive and liberatory potentials within political and economic structures. The project is related to normative goals of peace, social and environmental justice, and ecological well-being across species.

Hence, Kahn seeks to transcend the limited framework of environmental education and to radicalize contemporary demands for sustainable development. He envisions a critical ecopedagogy that calls for analysis of ecological crisis and sustainable development to be mandated across the curriculum, that entire schools and communities come to focus on the problem of sustainability in all its myriad aspects, unlike present educational standards or policies. Yet he is wary of a too uncritical perception of the concept of "sustainable development" as a panacea to crisis since the concept itself is both nebulous and presently being utilized by all manner of corporations and states to legitimate ecologically unsustainable forms of globalization and imperialism.

Kahn is thus sketching out a project that requires further development, debate, and new concepts and teaching strategies as we learn more about the environment, ecological crisis, and ways we can develop a more sustainable lifestyle and ways of living on the planet. It could be that the current global economic crisis, in conjunction with growing ecological crisis, will force us to rethink the consumer society and our drive to always create more and bigger technologies and cities and to celebrate high-consumption and high-tech lifestyles. Likewise, the global energy crisis could force us to produce new energy technologies and modes of transportation and habitation that are more ecologically sound. Or, more provocatively, it may require us to reconstruct educational emphases on the "new" and "improved" so that society can more effectively evaluate and adopt past options that became unfortunately outmoded through the unceasing drive for hegemonic forms of progress.

Currently, educational, environmental, and economic policies are up for grabs in the United States and globally, as the political class and citizens grope with tremendous socioeconomic, environmental, and existential crises. The era of neoliberalism, based on a market fundamentalism that saw corporate laissez-faire solutions as the key to all social problems and economic development, is certainly ending but it is not yet clear what policies and philosophies will replace it. What follows could be worse still. In this uncertain situation, it is up to critical educators and concerned citizens to re-envision the importance of education as a means through which we can engage our current set of crises, as we develop pedagogies adequate to the challenges of the contemporary moment that can promote social transformation guided by concerns of sustainability and justice. Richard Kahn has produced an important pedagogical intervention into the ever-mounting global ecological crisis and offered critical perspectives on ways that ecopedagogy and ecoliteracy can be developed as palpable alternatives to the status

status quo. It is important now for others to take up this task and to make critical ecopedagogy an important component of the reconstruction of education and society.

Professor Douglas Kellner
George F. Kneller Philosophy of Education Chair
University of California, Los Angeles

Bibliography

Abbey, E. 2000. *The Monkeywrench Gang*. New York: Harper Perennial Modern Classics.

Abromeit, J., & W. M. Cobb. (Eds.). 2004. *Herbert Marcuse: A Critical Reader*. New York: Routledge.

Adamson, J., M. M. Evans, & R. Stein. 2002. *The Environmental Justice Reader: Politics, Poetics, & Pedagogy*. Tucson: University of Arizona Press.

Adler, M. J. 1982. *The Paideia Proposal: An Educational Manifesto*. New York: Macmillan Publishing.

Adorno, T. 2000. *The Adorno Reader*. Edited by B. O'Connor. Oxford, UK: Blackwell Publishers.

Adorno, T., & M. Horkheimer, 1979. *Dialectic of Enlightenment*. London: Verso.

Agamben, G. 2004. *The Open: Man and Animal*. Stanford: Stanford University Press.

Ahlberg, M. 1998. Ecopedagogy and Ecodidactics: Education for Sustainable Development, Good Environment and Good Life. *Bulletins of the Faculty of Education* 69. University of Joensuu.

Aldred, L. 2000. Plastic Shamans and Astroturf Sun Dances: New Age Commercialization of Native American Spirituality. *The American Indian Quarterly* 24(3): 329–52.

Allen, R. L. 2001. The Globalization of White Supremacy: Toward a Critical Discourse on the Racialization of the World. *Educational Theory* 51(4): 467–85.

Anderson, P. 2000. Renewals. *New Left Review* 11(1): 11.

Andrzejewski, J., M. Baltodano, & L. Symcox. (Eds.). 2009. *Social Justice, Peace, and Environmental Education: Transformative Standards*. New York: Routledge.

Antunes, A., & M. Gadotti. 2005. Eco-pedagogy as the Appropriate Pedagogy to the Earth Charter Process. In P. Blaze Corcoran (Ed.), *The Earth Charter in Action: Toward a Sustainable World*. Amsterdam: KIT Publishers.

Apple, M. 1992. Is New Technology Part of the Solution or Part of the Problem in Education. In J. Beynon & H. Mackay (Eds.), *Technological Literacy and the Curriculum*. London: Falmer

Press: 105–24.

Archie, M. L. 2003. *Advancing Education Through Environmental Literacy*. Alexandria, VA: Association for Supervision and Curriculum Development.

Aristotle. 1943. *Politics*.Trans. Benjamin Jowett. New York: Random House.

Aronowitz, S. 1985. Why Should Johnny Read? *The Village Voice Literary Supplement* (May): 13.

Bahro, R. 1994. *Avoiding Social and Ecological Disaster: The Politics of World Transformation*. Bath, U.K.: Gateway Books.

Bantock, G. H. 1980. *Studies in the History of Educational Theory: Volume 1, Artifice and Nature, 1350–1765*. London: George Allen & Unwin LTD.

Bari, J. 1997. Revolutionary Ecology. *Capitalism, Nature, Socialism: A Journal of Socialist Ecology* 8(2): 145–49.

Barnhardt, R., & A. O. Kawagley. 2005. Indigenous Knowledge Systems and Native Alaska Ways of Knowing. *Anthropology and Education Quarterly* 36(1): 8–23.

Bee, B. 1981. The Politics of Literacy. In R. Mackie (Ed.), *Literacy & Revolution: The Pedagogy of Paulo Freire*. New York: Continuum.

Berkes, F. 1993. Traditional Ecological Knowledge in Perspective. In J. T. Inglis (Ed.), *Traditional Ecological Knowledge: Concepts and Cases*. Ottawa: International Development Research Centre.

Berry, T. 1999. *The Great Work: Our Way into the Future*. New York: Bell Tower.

———. 1988. *The Dream of the Earth*. San Francisco: Sierra Club Books.

Besser, H. 1993. Education as Marketplace. In R. Muffoletto & N. Knupfer (Eds.), *Computers in Education: Social, Historical, and Political Perspectives*. Cresskill, NJ: Hampton Press.

Best, S. 2003. Common Natures, Shared Fates: Toward an Interspecies Alliance Politics. *Impact Press* (December/January). Online at: http://www.impactpress.com/articles/decjan03/interspecies12103.html.

Best, S., & R. Kahn. 2005. Trial by Fire: The SHAC7, Globalization, and the Future of Democracy. *Animal Liberation Philosophy and Policy Journal* 2(2). Online at: http://www.cala-online.org/Journal/Issue3/Trial%20by%20Fire.htm.

Best, S., & D. Kellner. 2001. *The Postmodern Adventure: Science, Technology and Cultural Studies at the Third Millennium*. New York: Guilford.

Best, S. & A. J. Nocella, II. 2006. A Fire in the Belly of the Beast: The Emergence of Revolutionary Environmentalism. In S. Best & A. J. Nocella, II (Eds.), *Igniting a Revolution: Voices in Defense of the Earth*. Oakland: AK Press.

———. 2004. *Terrorists or Freedom Fighters? Reflections on the Liberation of Animals*. New York: Lantern Press.

Bey, H. 1985. *The Temporary Autonomous Zone, Ontological Anarchy, Poetic Terrorism*. New York: Autonomedia.

Blanke, H. T. 1996. Domination and Utopia: Marcuse's Discourse on Nature, Psyche, and Culture. In D. Macauley (Ed.), *Minding Nature: The Philosophers of Ecology*. New York: Guilford Press: 186–208.

Boff, L. 2008. *Essential Care: An Ethics of Human Nature*. Waco, TX: Baylor University Press.

———. 1997. *Cry of the Earth, Cry of the Poor*. Maryknoll, NY: Orbis Books.

Bookchin, M. 1982. *The Ecology of Freedom: The Emergence and Dissolution of Hierarchy*. Palo Alto, CA: Chesire Books.

Bookchin, M., & D. Foreman. 1991. *Defending the Earth: A Dialogue between Murray Bookchin and Dave Foreman*. Edited by Steve Chase. Boston: South End Press.

Bowers, C. A. 2006a. *Transforming Environmental Education: Making the Renewal of the Cultural and Environmental Commons the Focus of Educational Reform*. EcoJustice Press. Online at: https://scholarsbank.uoregon.edu/dspace/bitstream/1794/3070/6/transEE-rev.pdf.

———. 2006b. *Revitalizing the Commons: Cultural and Educational Sites of Resistance and Affirmation*. Lanham, MD: Lexington Books.

———. 2003a. Can Critical Pedagogy Be Greened? *Educational Studies* 34: 11–21.

———. 2003b. The Environmental Ethic Implicit in Three Theories of Evolution. *The Trumpeter* 19(3): 67–86.

———. 2001. *Educating for Eco-Justice and Community*. Athens: University of Georgia Press.

———. 2000. *Let Them Eat Data: How Computers Affect Education, Cultural Diversity, and the Prospects of Ecological Sustainability*. Athens: University of Georgia Press.

Bowers, C. A., and F. Apffel-Marglin. (Eds.) 2005. *Rethinking Freire: Globalization and the Environmental Crisis*. Hillsdale, NJ: Lawrence Erlbaum.

Brandt, W. 1980. *North-South: A Program for Survival*. Cambridge, MA: MIT Press.

Brechin, S. R., & D. A. Freeman. 2004. Public Support for Both the Environment and an Anti-Environmental President: Possible Explanations for the George W. Bush Anomaly. *The Forum* 2(1), Article 6.

Brickhouse, N. W., & J. M. Kittleson. 2006. Visions of Curriculum, Community, and Science. *Educational Theory* 56(2): 191–204.

Brinkley, D. 2009. Bob Dylan's America. *Rolling Stone*. April 29.

Bromley, H., & M. Apple. (Eds.). 1998. *Education/Technology/Power: Educational Computing as Social Practice*. Albany: State University of New York Press.

Broswimmer, F. J. 2002. *Ecocide: A Short History of the Mass Extinction of Species*. London: Pluto Press.

Brown, L. R. 2008. *Plan B 3.0: Mobilizing to Save Civilization* (3rd ed.). New York: W. W. Norton.

Brundtland, G. H., et al. 1987. *Our Common Future: Report of the World Commission on Environment and Development*. Oxford: Oxford University Press.

Burbules, N., & T. Callister. 2000. *Watch IT: The Risks and Promises of Information Technology*. Boulder, CO: Westview Press.

———. 1996. Knowledge at the Crossroads. *Educational Theory* 46(1): 23–34.

Butts, R. F. 1973. *The Education of the West: A Formative Chapter in the History of Civilization*. New York: McGraw-Hill.

Cajete, G. 2000. *Native Science: Natural Laws of Interdependence*. Santa Fe, NM: Clear Light Publishers.

———. 1999a. *Igniting the Sparkle: An Indigenous Science Education Model*. Skyland, NC: Kivaki Press.

———. 1999b. Reclaiming Biophilia: Lessons from Indigenous Peoples. In R. Williams (Ed.), *Ecological Education in Action: On Weaving Education, Culture, and the Environment*. Albany: SUNY Press.

———. 1993. *Look to the Mountain*. Skyland, NC: Kivaki Press.

Calderon, D. 2006. Developing Critical Interstitial Methodology: Taking Greater Control Over Our Resistance. In B. Kozuh, R. Kahn, A. Kozlowska, and P. Krope (Eds.), *Description and Explanation in Educational and Social Research*. Los Angeles and Warsaw: Rodn "WOM" Publishers.

Camara, H. 1995. *Sister Earth: Creation Ecology & The Spirit*. New York: New City Press.

Capra, F. 2002. *The Hidden Connections: Integrating the Biological, Cognitive, and Social Dimensions of Life into a Science of Sustainability*. New York: Doubleday.

———. 2000. Ecoliteracy: A Systems Approach to Education. In M. Crabtree (Ed.), *Ecoliteracy: Mapping the Terrain*. Berkeley: Learning in the Real World.

———. 1996. *The Web of Life: A New Scientific Understanding of Living Systems*. New York: Anchor Books.

———. 1984. *The Turning Point: Science, Society and the Rising Culture*. New York: Bantam Books.

Castells, M. (1999). Flows, Networks, Identities: A Critical Theory of the Information Society. In M. Castells, R. Flecha, P. Freire, H. A. Giroux, D. Macedo, & P. Willis (Eds.), *Critical Education in the New Information Age*. Lanham, MD: Rowman & Littlefield.

———. 1996. *The Information Age: Economy, Society and Culture Vol. I: The Rise of the Network Society*. Cambridge: MA. Blackwell Publishers.

Cayley, D. 2005. *The Rivers North of the Future: The Testament of Ivan Illich*. Toronto, ON: House of Anansi Press.

———. 1992. *Ivan Illich in Conversation*. Concord, Ontario: House of Anansi Press.

Chalk, P., B. Hoffman, R. Reville, & A.-B. Kasupski. 2005. Trends in Terrorism: Threats to the United States and the Future of the Terrorism Risk Insurance Act (Santa Monica: RAND Corporation). Online at: http://rand.org/pubs/monographs/2005/ RAND_MG393.pdf.

Churchill, W. 1996. *Fantasies of the Master Race*. Monroe, ME: Common Courage Press.

Clark, J. L. 1985. Thus Spoke Chief Seattle: The Story of an Undocumented Speech, *Prologue: The Journal of the National Archives*, 17(58): 58–65.

Clarke, K., & J. Hemphill. 2001. *The Santa Barbara Oil Spill: A Retrospective*. In Proceedings of the 64th Annual Meeting of the Association of Pacific Coast Geographers, UC Santa Barbara. Online at: http://www.geog.ucsb.edu/~jeff/sb_69oilspill/santa_barbara_1969_oilspill.pdf.

Cleaver, E. (Ed.). 1970. *Ramparts*, Ecology Special. *Ramparts* Magazine. May issue.

Clinton, W. J. 2000. Remarks by the President to the Community of the University of Warwick. December 14, Coventry, Warwickshire, England.

Cobern, W. W., & C. C. Loving. 2001. Defining "Science" in a Multicultural World: Implications for Science Education. *Science Education*, 85(1): 50–67.

Cohen, D. 1998. *Radical Heroes: Gramsci, Freire, and the Politics of Adult Education*. New York: Garland.

Collins, P. H. 1993. Toward a New Vision: Race, Class, and Gender as Categories of Analysis and Connection. *Race, Sex, & Class* 1: 25–45.

Cope, B., & M. Kalantzis (Eds.). 2000. *Multiliteracies: Literacy Learning and the Design of Social Futures*. New York: Routledge.

Corwin, M. 1989. The Oil Spill Heard 'Round the Country! *Los Angeles Times* (January 28, 1989). Online at: http://www.geog.ucsb.edu/~jeff/sb_69oilspill/69oilspill_articles.html.

Cosgrove, D. (2001). *Apollo's Eye: A Cartographic Genealogy of the Earth in the Western Imagination*. Baltimore: Johns Hopkins University Press.

Counts. G. S. 1932. *Dare the School Build a New Social Order?* New York: John Day Company.

Courts, P. L. 1998. *Multicultural Literacies: Dialect, Discourses, and Diversity*. New York: Peter Lang.

Covarrubias, A., & K. R. Weiss. 2005. Mysterious Oil Patches Take Big Toll on Seabirds. *Los Angeles Times*, January 21: 1.

Coyle, K. 2005. *Environmental Literacy in America. What 10 Years of NEETF/Roper Research and Related Studies Say About Environmental Literacy in the U.S.* Washington, DC: National Environmental Education & Training Foundation.

Cremin, L. A. 1980. *American Education: The National Experience 1783–1876*. New York: Harper

& Row.

Darder, A. 2002. *Reinventing Paulo Freire: A Pedagogy of Love*. Boulder, CO: Westview Press.

Darder, A., M. P. Baltodano, & R. Torres. (Eds.). 2008. *The Critical Pedagogy Reader* (2nd Ed.). New York: Routledge.

Delgado, S. L. 2005. Ecopedagogia y Cultura Depredadora. *Revista Cubana de Educación Superior* 2(1): 59–70.

Deloria, V., Jr., & D. R. Wildcat. 2001. *Power and Place: Indian Education in America*. Golden, CO: Fulcrum Resources.

DeLuca, K. 2002. The Frankfurt School and the Domination of Nature: New Grounds for Radical Environmentalism. In J. Nealon and C. Irr (Eds.), *Rethinking the Frankfurt School: Alternative Legacies of Cultural Critique*. Albany: SUNY Press.

Dewey, J. 1997 [1916]. *Democracy and Education: An Introduction to the Philosophy of Education*. Carbondale and Edwardsville: Southern Illinois University Press.

———. 1897. My Pedagogic Creed. *The School Journal* 54(3): 77–80.

Diamond, J. 2005. *Collapse: How Societies Choose to Fail or Succeed*. New York: Penguin.

Dowie, M. 1996. *Losing Ground: American Environmentalism at the Close of the Twentieth Century*. Cambridge, MA: MIT Press.

Dugger, C. 2006. Clinton, Impresario of Philanthropy, Gets a Progress Update. *New York Times*, April 1.

Dyrenfurth, M. J. 1991. Technological Literacy Synthesized. In M. J. Dyrenfurth & M. R. Kozak (Eds.), *Technological Literacy*. Peoria, IL: Glencoe, McGraw-Hill: 138–186.

Eisler, R. 2000. *A Blueprint for Partnership Education in the 21st Century*. Boulder, CO: Westview Press.

Eisler, R. & R. Miller. (Eds.). 2004. *Educating for a Culture of Peace*. Portsmouth, NH: Heinemann.

Eryaman, M. Y. 2009. *Peter McLaren, Education, and the Struggle for Liberation: Revolution as Education*. Cresskill, NJ: Hampton Press.

Espada, M. 2000. Viva Vieques! *The Progressive*, July 28: 27–29.

Eyerman, R., & A. Jamison. 1991. *Social Movements: A Cognitive Approach*. University Park, PA: Pennsylvania State University Press.

Fawcett, L., A. Bell, & C. Russell. 2002. Guiding Our Environmental Praxis: Teaching for Social and Environmental Justice. In W. Leal Filho (Ed.), *Teaching Sustainability at Universities: Towards Curriculum Greening*. New York: Peter Lang.

Fay, B. 1996. *Contemporary Philosophy of Social Science: A Multicultural Approach*. Malden, MA: Blackwell Publishing.

Feenberg, A. 2002. *Transforming Technology: A Critical Theory Revisited*. Oxford, UK: Oxford University Press.

——. 1999. *Questioning Technology*. New York and London: Routledge.

——. 1995. *Alternative Modernity*. Berkeley: University of California Press.

——. 1991. *Critical Theory of Technology*. New York: Oxford University Press.

Figueroa, R., & S. Harding. 2003. *Science and Other Cultures: Issues in Philosophies of Science and Technology*. New York: Routledge.

Flannery, T. 2006. *The Weather Makers: How Man Is Changing the Climate and What It Means for Life on Earth*. New York: Atlantic Monthly Press.

Foreman, D., & B. Haywood. (Eds.). 2002. *Ecodefense: A Field Guide to Monkeywrenching*. Third Edition. Chico, CA: Abbzug Press.

Foster, J. B. 2002. *Ecology Against Capitalism*. New York: Monthly Review Press.

Fotopolous, T. 1997. *Towards an Inclusive Democracy: The Crisis of the Growth Economy and the Need for a New Liberatory Project*. London: Cassell.

——. 1995. Direct and Economic Democracy in Ancient Athens and its Significance Today. *Democracy & Nature* 1(1) (Fall).

Fouts, R. 1997. *Next of Kin: My Conversations with Chimpanzees*. New York: Avon Books.

Freire, P. 2004. *Pedagogy of Indignation*. Boulder, CO: Paradigm Publishers.

——. 2001. *Pedagogy of the Oppressed*. New York: Continuum.

——. 2000. *Cultural Action for Freedom*. Cambridge, MA: Harvard Educational Review.

——. 1998a. *Pedagogy of Freedom: Ethics, Democracy and Civic Courage*. Lanham, MD: Rowman & Littlefield.

——. 1998b. *A Paulo Freire Reader*. New York: Herder and Herder.

——. 1997a. A Response. In P. Freire, J. W. Fraser, D. Macedo, T. McKinnon, & W. T. Stokes (Eds.), *Mentoring the Mentor: A Critical Dialogue with Paulo Freire*. New York: Peter Lang.

——. 1997b. *Pedagogy of the Heart*. New York: Continuum.

——. 1996. *Letters to Cristina: Reflections on My Life and Work*. New York: Routledge.

——. 1993. *Pedagogy of the City*. New York: Continuum.

——. 1992. *Pedagogy of Hope: Reliving the Pedagogy of the Oppressed*. New York: Continuum.

——. 1976. *Education: The Practice of Freedom*. London: Writers and Readers.

—— 1973. *Education for Critical Consciousness*. New York: Continuum.

———. 1972. *Pedagogy of the Oppressed*. New York: Herder & Herder.

Freire, P., & R. Davis. 1981. Education for Awareness: A Talk with Paulo Freire. In R. Mackie (Ed.), *Literacy & Revolution: The Pedagogy of Paulo Freire*. New York: Continuum.

Freire, P., & D. Macedo. 1987. *Literacy: Reading the Word and the World*. Westport, CT: Bergin & Garvey.

Gabbard, D. A. 1993. *Silencing Ivan Illich: A Foucauldian Analysis of Intellectual Exclusion*. Lanham, MD: Rowman & Littlefield.

Gadotti, M. 2009. *Education for Sustainability: A Contribution to the Decade of Education for Sustainable Development*. Sao Paulo, Brazil: Editora e Livraria Instituto Paulo Freire.

———. 2008. Paulo Freire and the Culture of Justice and Peace: The Perspective of Washington vs. The Perspective of Angicos. In C. Torres & P. Noguera (Eds.), *Social Justice Education for Teachers: Paulo Freire and the Possible Dream*. Denmark: Sense Publishers.

———. 2003. Pedagogy of the Earth and Culture of Sustainability. Paper presented at Lifelong Learning, Participatory Democracy and Social Change: Local and Global Perspectives conference, Toronto, Canada.

———. 2000. *Pedagogia da Terra*. Sao Paulo, Brazil: Peiropolis.

———. 1994. *Reading Paulo Freire*. Albany: SUNY Press.

Gault-Williams, M. 1987. *Don't Bank on Amerika: The History of the Isla Vista Riots of 1970*. Santa Barbara: Gault-Williams.

Giroux, H. 2006. *Beyond the Spectacle of Terrorism: Global Uncertainty and the Challenge of the New Media*. Boulder, CO: Paradigm Publishers.

———. 2000. *Stealing Innocence: Corporate Culture's War on Youth*. New York: Palgrave.

———. 1985. Introduction. In P. Freire, *The Politics of Education*. Westport, CT: Bergin & Garvey.

Glenn, J. L. 2000. *Environment-Based Education: Creating High Performance Schools and Students*. Washington, DC: National Environmental Education and Training Foundation.

Global Scenario Group. 2002. *Great Transition: The Promise and Lure of the Times Ahead*. Boston, MA: Stockholm Environment Institute.

Goldman, E. 1912. The Social Importance of the Modern School. Emma Goldman Papers. Rare Books and Manuscripts Division, New York Public Library.

González-Gaudiano, E. 2005. Education for Sustainable Development: Configuration and Meaning. *Policy Futures in Education* 3(3): 243–50.

Gore, A. 2006. *An Inconvenient Truth*. New York: Rodale Books.

———. 1994. Remarks Prepared for Delivery. Speech at the International Telecommunications Union (Buenos Aires). Online at: http://www.itu.int/itudoc/itu-d/wtdc/wtdc1994/speech/gore_ww2.doc.

Gossage J. P., L. Barton, L. Foster, L. Etsitty, C. LoneTree, C. Leonard, & P. A. May. 2003. Sweat Lodge Ceremonies for Jail-Based Treatment. *Journal of Psychoactive Drugs* 35(1): 33–42.

Gottleib, R. 1994. *Forcing the Spring: The Transformation of the American Environmental Movement*. San Francisco: Island Press.

Gough, A. G. 1993. *Founders in Environmental Education*. Geelong, Australia: Deakin University Press.

Grande, S. 2004. *Red Pedagogy: Native American Social and Political Thought*. Lanham, MD: Rowman & Littlefield.

Gray-Donald, J., & D. Selby. (Eds.). 2008. *Green Frontiers: Environmental Educators Dancing Away from the Mechanism*. Rotterdam, Netherlands: Sense Publishers.

Gronemeyer, M. 1987. Ecological Education a Failing Practice? Or: Is the Ecological Movement an Educational Movement? In W. Lierman & J. Kulich (Eds.), *Adult Education and the Challenge of the 1990s*. London: Croom Helm.

Grossman, E. 2004. High-Tech Wasteland. *Orion*. July/August. Online at: http://www.orion online.org/pages/om/04-4om/Grossman.html.

Grubb, W. N. 1996. The New Vocationalism—What It Is, What It Could Be. *Phi Delta Kappan* 77(8): 535–46.

Gruenewald, D. A. 2004. A Foucauldian Analysis of Environmental Education: Toward the Socioecological Challenge of the Earth Charter. *Curriculum Inquiry* 34(1): 71–107.

——. 2003. The Best of Both Worlds: A Critical Pedagogy of Place. *Educational Researcher* 32(4): 3–12.

Gruenewald, D. A., & G. Smith. (Eds.). 2007. *Place-Based Education in a Global Age: Local Diversity*. New York: Taylor & Francis.

Guillermo, K. 2005. Response to Nathan Snaza's (Im)possible Witness: Viewing PETA's "Holocaust on Your Plate." *Animal Liberation Philosophy and Policy Journal*. 2(1). Online at: http://www.cala-online.org/Journal/Issue3/ Response_Letter_Snaza.htm.

Gur-Ze'ev, I. (Ed.). 2005. *Critical Theory and Critical Pedagogy Today: Toward a New Critical Language in Education*. Haifa: Haifa University.

——. 2002. Bildung and Critical Theory in Face of Postmodern Education. *Journal of Philosophy of Education* 36(3): 391–408.

——. 1998. Toward a Nonrepressive Critical Pedagogy. *Educational Theory* 48(4): 463–86.

Gutierrez, F., & C. Prado. 1999. *Ecopedagogia e Cidadania Planetaria*. Sao Paulo, Brazil: Cortez.

Gwynn, A. 1966. *Roman Education from Cicero to Quintillian*. New York: Teachers College Press.

Habermas, J. 1984. *The Theory of Communicative Action, Vol 1: Reason and the Rationalization of*

Society. Boston: Beacon Press.

——. 1972. *Knowledge & Human Interest*. Boston: Beacon Press.

Hadas, M. 1959. *Hellenistic Culture: Fusion and Diffusion*. New York: Columbia University Press.

Hall, R. 1985. Distribution of the Sweat Lodge in Alcohol Treatment Programs. *Current Anthropology* 26(1): 134–35.

Hall, S. 1987. Gramsci and Us. *Marxism Today* (June): 19.

Haluza-DeLay, R. 2006. *Developing a Compassionate Sense of Place: Environmental and Social Conscientization in Environmental Organizations*. Unpublished Doctoral Dissertation, University of Western Ontario. Online at: http://csopconsulting.tripod.com/sitebuilder content/sitebuilderfiles/Haluza-DeLay-FINAL_Diss.pdf.

Hammer, R. 2006. Teaching Critical Media Literacies: Theory, Praxis and Empowerment. *InterActions: UCLA Journal of Education and Information Studies* 2(1), Article 6. Online at: http://repositories.cdlib.org/gseis/interactions/vol2/iss1/6.

——. 1995. Strategies for Media Literacy. In P. McLaren, R. Hammer, D. Sholle, & S. Reilly. *Rethinking Media Literacy: A Critical Pedagogy of Representation*. New York: Peter Lang: 225–35.

Hammer, R., & D. Kellner. (Eds.). 2009. *Media/Cultural Studies: Critical Approaches*. New York: Peter Lang.

——. 2001. Multimedia pedagogy and multicultural education for the new millennium. *Current Issues in Education* 4(2). Online at: http://cie.ed.asu.edu/volume4/number2/.

Haraway, D. 2003. *The Companion Species Manifesto: Dogs, Species, and Significant Otherness*. Chicago: Prickly Paradigm Press.

——. 1988. Situated Knowledges: The Science Question in Feminism and the Privilege of Partial Perspective. *Feminist Studies* 14(3): 575–99.

Hardin, G. 1968. The Tragedy of the Commons. *Science* 162: 1243–48.

Harding, S. 2008. *Sciences From Below: Feminisms, Postcolonialities, and Modernities*. Durham, NC: Duke University Press.

——. (Ed.). 2004a. *The Feminist Standpoint Theory Reader: Intellectual and Political Controversies*. New York and London: Routledge.

——. 2004b. How Standpoint Methodology Informs Philosophy of Social Science. In S. N. Hesse-Biber & P. Leavy (Eds.), *Approaches to Qualitative Research: A Reader on Theory and Practice*. New York: Oxford University Press.

——. 1998. *Is Science Multicultural?: Postcolonialisms, Feminisms, and Epistemologies*. Bloomington: Indiana University Press.

——. 1993. *The "Racial" Economy of Science: Toward a Democratic Future*. Bloomington: Indiana University Press.

———. 1991. *Whose Science? Whose Knowledge?* Ithaca, NY: Cornell University Press.

Hardison, P. 2006. Indigenous Peoples and the Commons. Nov. 20. Online at: http://www.onthecommons.org/content.php?id=962.

Hardt, M., & A. Negri. 2004. *Multitude: War and Democracy in the Age of Empire.* New York: Routledge.

———. 2000. *Empire.* Cambridge, MA: Harvard University Press.

Harkin, M. E., & D. R. Lewis. (Eds.). 2007. *Native Americans and the Environment: Perspectives on the Ecological Indian.* Lincoln: University of Nebraska Press.

Harney, C. 1995. *The Way It Is.* Nevada City, CA: Blue Dolphin Press.

Havelock, E. A. 1986. *The Muse Learns to Write: Reflections on Orality and Literacy from Antiquity to Present.* New Haven, CT: Yale University Press.

Hay, C. 1999. Crisis and the Structural Transformation of the State: Interrogating the Process of Change. *British Journal of Politics and International Relations* 1(3): 317–44.

Hayden, M. 1989. What Is Technological Literacy? *Bulletin of Science, Technology and Society* 119: 220–33.

Heinonen, S., P. Jokinen, & J. Kaivo-oja. 2001. The Ecological Transparency of the Information Society. *Future* 33: 319–37.

Hickman, L. 2001. *Philosophical Tools for Technological Culture.* Bloomington: Indiana University Press.

Hill, D., & S. Boxley. 2007. Critical Teacher Education for Economic, Environmental and Social Justice: An Ecosocialist Manifesto. *Journal for Critical Education Policy Studies* 5(2). Online at: http://www.jceps.com/?pageID=article&articleID=96.

Hill, L., & D. Clover. 2003. *Environmental Adult Education: Ecological Learning, Theory, and Practice for Socioenvironmental Change.* San Francisco: Jossey-Bass.

Hoinacki, L., & C. Mitcham. (Eds.). 2002. *The Challenges of Ivan Illich: A Collective Reflection.* Albany: SUNY Press.

Holbrook, J., A. Mukherjee, & V. S. Varma. (Eds.). 2000. *Scientific and Technological Literacy for All.* UNESCO and International Council of Associations for Science Education. Delhi, India: Center for Science Education and Communication.

hooks, b. 2009. *Belonging: A Culture of Place.* New York: Taylor & Francis.

Horkheimer, M., & T. Adorno. 2002. *Dialectic of Enlightenment: Philosophical Fragments.* Palo Alto, CA: Stanford University Press.

Hyslop-Margison, E. J., & M. A. Naseem. 2007. *Scientism and Education: Empirical Research as Neo-Liberal Ideology.* New York: Springer.

Illich, I. 1995. Statements by Jacques Ellul and Ivan Illich. *Technology in Society* 17(2): 231–38.

———. 1992a. *In the Mirror of the Past: Lectures and Addresses 1978–1990*. New York: Marion Boyars.

———. 1992b. *In the Vineyard of the Text: A Commentary to Hugh's Didasacalicon*. Chicago: University of Chicago Press.

———. 1988. *Alternativas II*. Mexico: Joaquín Mortiz / Planeta.

———. 1982. *Gender*. New York: Pantheon.

———. 1978. *The Right to Useful Unemployment and Its Professional Enemies*. London: Marion Boyars.

———. 1973. *Tools for Conviviality*. New York: Harper and Row.

———. 1971. *Celebration of Awareness*. New York and London: Marion Boyars.

———. 1970. *Deschooling Society*. New York and London: Marion Boyars.

Illich, I., & E. Verne. 1981. *Imprisoned in the Global Classroom*. London: Writers & Readers.

Ince, J. F. 1995. *The Earth Pledge Book: A Call for Commitment*. Sausalito, CA: Timely Visions Publishing Company.

Jaeger, W. 1945. *Paideia: The Ideals of Greek Culture*. New York: Oxford University Press.

Jarboe, J. F. 2002. The Threat of Eco-Terrorism. Testimony before the House Resources Committee, Subcommittee on Forests and Forest Health, February 12, 2002. Online at: http://www.fbi.gov/congress/congress02/jarboe021202.htm.

Jardine, D. W. 2000. *"Under the Tough Old Stars": Ecopedagogical Essays*. Brandon, VT: Solomon Press.

Jegede, O. 2002. An Integrated ICT-Support for ODL in Nigeria: The Vision, the Mission and the Journey so Far. Paper prepared for the LEARNTEC-UNESCO 2002 Global Forum on Learning Technology. Karlsruhe, Germany.

Jenkins, E. W. 1997. Technological Literacy: Concepts and Constructs. *Journal of Technology Studies* 23(1): 2–5.

Jensen, D. 2007. *Endgame, Vol. II: Resistance*. New York: Seven Stories Press.

———. 2006. *Endgame, Vol. I: The Problem of Civilization*. New York: Seven Stories Press.

Jensen, R. 2009. Let Us Find Our Prophetic Voices: Finding a Stubborn Hope to Live in a Dead Culture. *Counterpunch*. June 16. Online at: http://www.counterpunch.org/jensen06162009.html.

Jickling, B. 2005. Sustainable Development in a Globalizing World: A Few Cautions. *Policy Futures in Education* 3(3): 251–59.

Jickling, B., & A. E. Wals. 2007. Globalization and Environmental Education: Looking Beyond Sustainable Development. *Journal of Curriculum Studies* 40(1): 1–21.

Jucker, R. 2002. *Our Common Illiteracy: Education as If the Earth and People Mattered*. New York: Peter Lang.

Kahn, R. 2009. Anarchic Epimetheanism: The Pedagogy of Ivan Illich. In R. Amster, A. DeLeon, L. Fernandez, A. J. Nocella II, & D. Shannon (Eds.), *Contemporary Anarchist Studies: An Introductory Anthology of Anarchy in the Academy*. London: Routledge.

——. 2005. How the West Was One? The American Frontier and the Rise of a Global Internet Imaginary. *InterActions* 1(2). Online at: http://repositories.cdlib.org/gseis/interactions/vol1/iss2/art6.

——. 2003. Paulo Freire and Eco-Justice: Updating Pedagogy of the Oppressed for the Age of Ecological Calamity. *Freire Online Journal* 1(1). Online at: http://www.paulofreireinstitute.org/freireonline/volume1/1kahn1.html.

Kahn, R., & B. Humes. 2009. Marching out from Ultima Thule: Critical Counterstories of Emancipatory Educators Working at the Intersection of Human Rights, Animal Rights, and Planetary Sustainability. *Canadian Journal of Environmental Education* 14(1).

Kahn, R., & D. Kellner. 2008. Technopolitics, Blogs, and Emergent Media Ecologies: A Critical/Reconstructive Approach. *Small Tech: The Culture of Digital Tools*. In B. Hawk, D. Rieder, & O. Oviedo (Eds.), Minneapolis: University of Minnesota Press.

——. 2007. Globalization, Technopolitics, and Radical Democracy. In L. Dahlberg & E. Siapera (Eds.), *Radical Democracy and the Internet: Interrogating Theory and Practice*, London: Palgrave: 17–36.

——. 2006. Resisting Globalization. In G. Ritzer (Ed.), *The Blackwell Companion to Globalization*. Malden, MA: Blackwell.

——. 2005. Oppositional Politics and the Internet: A Critical/Reconstructive Approach. *Cultural Politics*, Vol. 1, No. 1. Berg Publishers.

Kahn, R., & A. J. Nocella, II. (Eds.). Forthcoming. *Greening the Academy: Environmental Studies in the Liberal Arts*. Syracuse, NY: Syracuse University Press.

Kazamias, A. M. 2000. Education and the Polity/State: The Education of the Citizen–From the Ancient Polis (City State) to the Modern Ethnopolis (Nation State) and the New Cosmopolis of Late Modernity. Paper presented at the Academy of Athens, Greece, June.

Kellner, D. 2006. Introduction: Marcuse's Challenges to Education. *Policy Futures in Education* 4(1).

——. (Ed.). 2005a. *Herbert Marcuse: The New Left and the 1960s, Collected Papers of Herbert Marcuse, Volume Three*. London and New York: Routledge.

——. 2005b. *Media Spectacle and the Crisis of Democracy: Terrorism, War, and Election Battles*. Boulder, CO: Paradigm Publishers.

——. 2004. Technological Transformation, Multiple Literacies, and the Re-Visioning of Education. *E-Learning* 1(1).

——. 2003a. *Media Spectacle*. London and New York: Routledge.

———. 2003b. *From 9/11 to Terror War: The Dangers of the Bush Legacy*. Lanham, MD: Rowman & Littlefield.

———. 2003c. Toward a Critical Theory of Education. *Democracy & Nature*, Vol. 9, No. 1. Taylor and Francis: 51–64.

———. 2002a. Technological Revolution, Multiple Literacies, and the Restructuring of Education. In I. Snyder (Ed.), *Silicon Literacies: Communication, Innovation and Education in the Electronic Age*. London: Routledge: 154–69.

———. 2002b. Theorizing Globalization. *Sociological Theory*, 20(3): 285–305.

———. (Ed.). 2001. *Towards a Critical Theory of Society: Collected Papers of Herbert Marcuse, Vol. II*. New York: Routledge.

———. 2000. Globalization and New Social Movements: Lessons for Critical Theory and Pedagogy. N. Burbules & C. A. Torres (Eds.), *Globalization and Education: Critical Perspectives*. New York: Routledge.

———. 1998. Multiple Literacies and Critical Pedagogy in a Multicultural Society. *Educational Theory* 48: 103–22.

———. 1995. *Media Culture: Identity and Politics Between the Modern and the Postmodern*. New York: Routledge.

———. 1992. Marcuse, Liberation and Radical Ecology. *Capitalism, Nature, Socialism* 3(3): 43–46.

———. 1989. *Critical Theory, Marxism and Modernity*. Baltimore: Johns Hopkins University Press.

———. 1984. *Herbert Marcuse and the Crisis of Marxism*. Berkeley: University of California Press.

Kellner, D., T. Lewis, & C. Pierce. 2008. *On Marcuse: Critique, Liberation, and Reschooling in the Radical Pedagogy of Herbert Marcuse*. Rotterdam, Netherlands: Sense Publishers.

Kellner, D., T. Lewis, C. Pierce, & D. Cho. (Eds.). 2008. *Marcuse's Challenge to Education*. Lanham, MD: Rowman & Littlefield.

Kellner, D., & J. Share. 2005. Toward Critical Media Literacy: Core Concepts, Debates, Organizations and Policies. *Discourse: Studies in the Cultural Politics of Education*. University of Queensland, Australia: Routledge.

Kelly, K. 1998. *New Rules for the New Economy*. London: Fourth Estate.

Kerenyi, K. 1976. *Dionysos: Archetypal Image of Indestructible Life*. Princeton: Princeton University Press.

Kincheloe, J. L. 2008. *Critical Pedagogy Primer*. New York: Peter Lang Publishers.

Klein, N. 2007. *The Shock Doctrine: The Rise of Disaster Capitalism*. New York: Metropolitan Books.

Kolbert, E. 2006. *Field Notes from a Catastrophe: Man, Nature, and Climate Change*. London: Bloomsbury Publishers.

Kovel, J. 2007. *The Enemy of Nature: The End of Capitalism or the End of the World?* (2nd Ed.). New York: Zed Books.

———. 1983. Theses on Technocracy. *Telos.* No. 54 (Winter).

Kress, G. 1997. Visual and Verbal Modes of Representation in Electronically Mediated Communication: The Potentials of New Forms of Text. In I. Snyder (Ed.), *Page to Screen: Taking Literacy into the Electronic Era.* Sydney, Australia: Allen & Unwin: 53–79.

Kuletz, V. 1998. *Tainted Desert: Environmental and Social Ruin in the American West.* New York: Routledge.

Kunstler, J. H. 2005. *The Long Emergency: Surviving the Converging Catastrophes of the Twenty-First Century.* New York: Atlantic Monthly Press.

LaCapra, D. 2009. *History and Its Limits: Human, Animal, Violence.* Ithaca, NY: Cornell University Press.

LaDuke, W. 2005. *Recovering the Sacred: The Power of Naming and Claiming.* Boston: South End Press.

Lankshear, C., & M. Knobel. 2000. Mapping Postmodern Literacies: A Preliminary Chart. *The Journal of Literacy and Technology.* Vol. 1, No. 1, Fall. Online at: http://www.literacy andtechnology.org/ v1n1/lk.html.

Lankshear, C., & I. Snyder. 2000. *Teachers and Technoliteracy: Managing Literacy, Technology and Learning in Schools.* Sydney, Australia: Allen & Unwin.

Lather, P. 2007. *Getting Lost: Feminist Efforts Toward a Double(d) Science.* Albany: SUNY Press.

Latour, B. 2004. *Politics of Nature.* Cambridge, MA: Harvard University Press.

Lenhart, A., J. Horrigan, L. Rainie, K. Allen, A. Boyce, M. Madden, & E. O'Grady. 2003. *The Ever-Shifting Internet Population: A New Look at Internet Access and the Digital Divide.* The Pew Internet & American Life Project. Online at: http://www.pewinternet.org/pdfs/ PIP_Shifting_Net_Pop _Report.pdf.

Lewis, J. 2005. Statement of John Lewis, Deputy Assistant Director, Federal Bureau of Investigation. Oversight on Eco-terrorism specifically examining the Earth Liberation Front ("ELF") and the Animal Liberation Front ("ALF"), U.S. Senate Committee on Environment & Public Works, May 18, 2005. Online at: http://epw. senate.gov/hearing_statements.cfm?id=237817.

Lewis, T. 2003. The Surveillance Economy of Post-Columbine Schools. *The Review of Education, Pedagogy & Cultural Studies,* 25(4): 335–56.

Lewis, T., & D. Cho. 2006. Home Is Where the Neurosis Is: A Topography of the Spatial Unconscious. *Cultural Critique.* Number 64: 69–91.

Lewis, T., & C. Gagel. 1992. Technological Literacy: A Critical Analysis. *Journal of Curriculum Studies* 24(2): 117–38.

Lewis, T., & R. Kahn. Forthcoming. *Education Out of Bounds: Cultural Studies for a Posthuman Age*. New York: Palgrave Macmillan.

Liddell, H. G., & R. Scott. 1940. *A Greek-English Lexicon*. Oxford: Clarendon Press.

Lomawaima, K. T., & T. McCarty. 2006. *To Remain an Indian: Lessons in Democracy from a Century of Native American Education*. New York: Teachers College Press.

Lonsdale, M., & D. McCurry. 2004. *Literacy in the New Millennium*. Adelaide, Australia: NCVER.

Lopez, B. 2007. The Leadership Imperative: An Interview with Oren Lyons. *Orion Magazine*. January/February. Online at: http://www.orionmagazine.org/index.php/articles/article/94/.

Lorde, A. 1990. Age, Race, Class and Sex: Women Redefining Difference. In R. Ferguson, et. al. (Eds.), *Out There: Marginalization and Contemporary Cultures*. Cambridge, MA: MIT Press: 281–88.

Luke, C. 2000. Cyber-Schooling and Technological Change: Multiliteracies for New Times. In B. Cope & M. Kalantzis (Eds.), *Multiliteracies: Literacy, Learning, and the Design of Social Futures*. South Yarra, Australia: Macmillan: 69–105.

———. 1997. *Technological Literacy*. Melbourne, Australia: National Languages & Literacy Institute. Adult Literacy Network.

Luke, T. W. 1999. *Capitalism, Democracy, and Ecology: Departing from Marx*. Urbana and Chicago: University of Illinois Press.

———. 1994. Marcuse and Ecology. In J. Bokina & T. J. Lukes (Eds.), *Marcuse: From the New Left to the Next Left*. Lawrence, Kansas: University of Kansas Press.

Lummis, G. 2002. Globalisation: Buidling a Partnership Ethic for an Ecopedagogy in Western Australia. *The Australian Journal of Teacher Education* 27(1).

Macedo, D., & S. R. Steinberg. (Eds.). 2007. *Media Literacy: A Reader*. New York: Peter Lang.

Magalhaes, H. G. D. 2005. Ecopedagogia y Utopia. *Educação Temática Digital, Campinas* 7(1): 53–60.

Malott, C. 2008. *A Call to Action: An Introduction to Education, Philosophy, and Native North America*. New York: Peter Lang.

Marcos, S. I. 2001. *Our Word Is Our Weapon: Selected Writings*. New York: Seven Stories Press.

Marcuse, H. 2001. The Individual in the Great Society. In D. Kellner (Ed.), *Towards a Critical Theory of Society*. New York and London: Routledge.

———. 1992. Ecology and the Critique of Modern Society. *Capitalism, Nature, Socialism* 3(3) (Sept.): 38–40.

———. 1979. Kinder des Prometheus. 25 Thesen zu Technik und Gesellschaft. *Neues Forum. Heft*

307/8 (August): Wien.

———. 1977. *The Aesthetic Dimension: Toward a Critique of Marxist Aesthetics*. Boston: Beacon Press.

———. 1972a. *Counter-Revolution and Revolt*. Boston: Beacon Press.

———. 1972b. Art in the One-Dimensional Society. In L. Baxandall (Ed.), *Radical Perspectives in the Arts*. Harmondsworth, UK: Penguin: 53–67.

———. 1972c. *An Essay on Liberation*. Harmondsworth, PA: Penguin.

———. 1969. *An Essay on Liberation*. Boston: Beacon.

———. 1968. Liberation from the Affluent Society. In D. Cooper (Ed.), *The Dialectics of Liberation*. Harmondsworth/Baltimore: Penguin: 175–92.

———. 1966. *Eros and Civilization*. Boston: Beacon Press.

———. 1965. Repressive Tolerance. In R. P. Wolff, B. Moore, & H. Marcuse (Eds.), *A Critique of Pure Tolerance*. Boston: Beacon Press.

———. 1964. *One-Dimensional Man: Studies in the Ideology of Advanced Industrial Society*. London: Routledge & Kegan Paul.

Marrou, H. I. 1964. *A History of Education in Antiquity*. New York: Mentor Press.

Martin, G. 2007. The Poverty of Critical Pedagogy: Toward a Politics of Engagement. In P. McLaren & J. L. Kincheloe (Eds.), *Critical Pedagogy: Where Are We Now?* New York: Peter Lang Publishers.

Martusewicz, R. 2005. Eros in the Commons: Educating for Eco-ethical Consciousness in a Poetics of Place. *Ethics, Place and Environment* 8(3): 341–48.

Martusewicz, R., & J. Edmundson. 2005. Social Foundations as Pedagogies of Responsbility and Eco-Ethical Commitment. In D. W. Butin (Ed.), *Teaching Social Foundations of Education: Contexts, Theories, and Issues*. Mahwah, NJ: Lawrence Erlbaum.

Marx, K. 1990. *Capital. Vol. 1*. Trans. B. Fowkes. London: Penguin Books.

Mason, J. 1998. *An Unnatural Order: Why We Are Destroying the Planet and Each Other*. New York: Continuum Press.

Matthews, M. 1994. *Science Teaching: The Role of History and Philosophy of Science*. New York: Routledge.

Mayo, P. 2001. Revolutionary Learning, Biodiversity, and Transformative Action. *Comparative Education Review* 45(1): 140–48.

McKenzie, M., P. Hart, H. Bai & B. Jickling. (Eds.). 2009. *Fields of Green: Restorying Culture, Environment, and Education*. Creskill, NJ: Hampton Press.

McLaren, P. 2000a. *Che Guevara, Paulo Freire, and the Pedagogy of Revolution*. Lanham, MD: Rowman & Littlefield.

———. 2000b. Paulo Freire's Pedagogy of Possibility. In S. F. Steiner, H. M. Krank, P. McLaren, & R. E. Bahruth (Eds.), *Freirian Pedagogy, Praxis, and Possibilities: Projects for the New Millennium*. New York: Taylor & Francis.

———. 1991. Schooling the Postmodern Body. In H. Giroux (ed.), *Postmodernism, Feminism, and Cultural Politics*. Albany: SUNY Press.

McLaren, P., & T. T. da Silva. 1993. Decentering Pedagogy: Critical Literacy, Resistance and the Politics of Memory. In P. McLaren & P. Leonard (Eds.), *Paulo Freire: A Critical Encounter*. New York: Taylor & Francis.

McLaren, P., & E. González-Gaudiano. 1995. Education and Globalization, An Environmental Perspective—An Interview with Edgar González-Gaudiano. *International Journal of Educational Reform* 4(1): 72–78.

McLaren, P., R. Hammer, D. Sholle, & S. Reilly. 1995. *Rethinking Media Literacy: a critical pedagogy of representation*. New York: Peter Lang.

McLaren, P., & D. Houston. 2005. Revolutionary Ecologies: Ecosocialism and Critical Pedagogy. In P. McLaren, *Capitalists & Conquerors: A Critical Pedagogy Against Empire*. Lanham, MD: Rowman & Littlefield: 166–88.

McLaren, P., & N. Jaramillo. 2007. *Pedagogy and Praxis in the Age of Empire: Towards a New Humanism*. Rotterdam, Netherlands: Sense Publishers.

McLaren, P., & J. L. Kincheloe. (Eds.). 2007. *Critical Pedagogy: Where Are We Now?* New York: Peter Lang Publishers.

McLuhan, M. 1964. *Understanding Media: The Extensions of Man*. New York: Signet Books.

Merchant, C. 1994. *Ecology: Key Concepts in Critical Theory*. Amherst, NY: Prometheus Books.

Millennium Ecosystem Assessment [MEA]. 2005. *Ecosystems and Human Well-Being: Synthesis*. Washington, DC: Island Press.

Miller, J. P. 2007. *The Holistic Curriculum* (2nd ed.). Toronto, ON: University of Toronto Press.

Miller, R. (Ed.). 1991. *New Directions in Education: Selections from Holistic Education Review*. Brandon, VT: Holistic Education Press.

Morris, D., & R. Morris. 1966. *Men and Apes*. London: Hutchinson Press.

Morrison, R. 1995. *Ecological Democracy*. Boston: South End Press.

Morrow, R. A., & C. A. Torres. 2002. *Reading Freire and Habermas: Critical Pedagogy and Transformative Social Change*. New York: Teachers College Press.

———. 1995. *Social Theory and Education: A Critique of Theories of Social and Cultural Reproduction*. Albany: SUNY Press.

Munro, D. A., & M. W. Holdgate (Eds.). 1991. *Caring for the Earth. A Strategy for Sustainable Living*. Gland, Switzerland: The World Conservation Union, United Nations Environment Programme, and World Wildlife Fund.

Nader, L. 1996. *Naked Science: Anthropological Inquiries into Boundaries, Power, and Knowledge*. New York: Routledge.

National Commission on Excellence in Education. 1983. *A Nation at Risk: The Imperative for Educational Reform*. Washington, DC: U.S. Government Printing Office.

National Telecommunications & Information Administration. 2002. *A Nation Online: How Americans Are Expanding Their Use of the Internet*. Online at: http://www.ntia.doc.gov/ntiahome/dn/nationonline_020502.htm.

New London Group. 1996. A Pedagogy of Multiliteracies: Designing Social Futures. *Harvard Educational Review* 66: 60–92.

Nietzsche, F. 1908. *Human, All Too Human: A Book for Free Spirits*. Chicago: Charles H. Kerr & Company.

——. 1990. *Twilight of the Idols or How to Philosophize with a Hammer*. New York: Penguin.

Nkrumah, K. 1964. *Consciencism: Philosophy and Ideology for Decolonization*. New York: Monthly Review Press.

Nocella, A. J. II, S. Best & P. McLaren. (Eds.). Forthcoming. *Academic Repression: Reflections from the Academic Industrial Complex*. Oakland, CA: AK Press.

North American Association for Environmental Education. 2000. *Excellence in Environmental Education: Guidelines for Learning (K-12)*. Washington, DC: NAAEE.

O'Cadiz, P., P. L. Wong, & C. A. Torres. 1998. *Education & Democracy: Paulo Freire, Social Movements and Educational Reform in Sao Paulo*. Boulder, CO: Westview Press.

Ochs, P. 1968. Joe Hill. *The War Is Over*. New York: Barricade Music.

Ohlinger, J. 1995. Critical Views of Paulo Freire's Work. Online at: http://www3.nl.edu/academics/cas/ace/resources/JohnOhliger_Insight1.cfm.

Olivera, O. 2004. *Cochabamba: Water War in Bolivia*. Cambridge, MA: South End Press.

Olsen, R., & D. Rejeski (Eds.). 2007. *Environmentalism and the Technologies of Tomorrow: Shaping the Next Industrial Revolution*. Washington, DC: Island Press.

Ong, W. 1982. *Orality and Literacy*. New York: Methuen Press.

Orr, D. 2004. *Earth in Mind: On Education, Environment, and the Human Prospect* (2nd Ed.). Washington, DC: Island Press.

——. 2002. *The Nature of Design: Ecology, Culture and Human Intention*. Oxford: Oxford University Press.

——. 1992. *Ecological Literacy: Education and the Transition to a Postmodern World*. Albany: SUNY Press.

O'Sullivan, E. 1999. *Transformative Learning: Educational Vision for the 21st Century*. London: Zed Books.

O'Sullivan, E., A. Morrell, & M. A. O'Connor. (Eds.). 2002. *Expanding the Boundaries of Transformative Learning*. New York: Palgrave.

O' Sullivan, E., & M. Taylor. 2004. *Learning Toward an Ecological Consciousness: Selected Transformative Practices*. New York: Palgrave.

Ó Tuathail, G., & D. McCormack. 1999. The Technoliteracy Challenge: Teaching Globalization Using the Internet. *Journal of Geography in Higher Education* 22: 347–61.

Pacific Research Institute. 1999. The California 1999 Index of Leading Environmental Indicators (San Francisco, CA). Online at: http://www.pacificresearch.org/pub/sab/enviro/99_enviroindex/caindex.pdf.

Papert, S. 2000. The Future of School. Online at: http://www.papert.org/articles/freire/freirePart1.html.

Parenti, M. 2003. *The Assassination of Julius Caesar: A People's History of Ancient Rome*. New York: The New Press.

Park, L. S-H., & D. N. Pellow. 2004. Racial Formation, Environmental Racism, and the Emergence of Silicon Valley. *Ethnicities* 4(3): 403–24.

Patel, R. 2008. *Stuffed and Starved: The Hidden Battle for the World Food System*. Hoboken, NJ: Melville House Publishing.

Payne, P. G. 2005. Growing Up Green. *Journal of the HEIA* 12(3): 2–12.

Pearson, G., & A. T. Young. 2002. *Technically Speaking: Why All Americans Need to Know More About Technology*. Washington, DC: National Academies Press.

Pena, D. G. 1998. *Chicano Culture, Ecology, Politics: Subversive Kin*. Tucson: University of Arizona Press.

Perlman, F. 1983. *Against His-Story, Against Leviathan! An Essay*. Detroit: Red and Black.

Petrina, S. 2000a. The Political Ecology of Design and Technology Education: An Inquiry into Methods. *International Journal of Technology and Design Education* 10(3): 207–37.

———. 2000b. The Politics of Technological Literacy. *International Journal of Technology and Design Education* 10(2): 181–206.

Pickering, L. J. 2002. *The Earth Liberation Front: 1997–2002*. South Wales, NY: Arissa Publications.

Pierce, C. 2007. Designing Intelligent Knowledge: Epistemological Faith and the Democratization of Science. *Educational Theory* 57(2): 123–40.

Plato. 1961. *Republic* (Trans. P. Shorey). In E. Hamilton and H. Cairns (Eds.) *Plato: Collected Dialogues*. Princeton, NJ: Princeton University Press.

Plepys, A. 2002. The Grey Side of ICT. *Environmental Impact Assessment Review*. Vol. 22: 509–23.

Plotnick, E. 1999. Information Literacy. ERIC Clearinghouse on Information and Technology, Syracuse University. ED-427777.

Popper, K. 1981. Science, Pseudo-Science, and Falsifiability. In R. D. Tweney, M. E. Doherty & C. R. Mynatt (Eds.), *Scientific Thinking*. New York: Columbia University Press: 92–99.

———. 1959. *The Logic of Scientific Discovery*. New York: Basic Books.

Porter, D., & D. Craig. 2004. The Third Way and the Third World: Poverty Reduction and Social Inclusion in the Rise of "Inclusive" Liberalism. *Review of International Political Economy* 11(2): 387–423.

Posner, R. A. 2004. *Catastrophe: Risk and Response*. Oxford: Oxford University Press.

Postman, N. 1992. *Technopoly: The Surrender of Culture to Technology*. New York: Random House.

———. 1985. *Amusing Ourselves to Death*. New York: Viking-Penguin.

Power, C. 1987. Science and Technology Towards Informed Citizenship. *Castme Journal* 7(3): 5–18.

Prakash, M. S., & G. Esteva. 2008. *Escaping Education: Living as Learning in Grassroots Cultures* (2nd Ed.). New York: Peter Lang.

Prakash, M. S., & D. Stuchul. 2004. McEducation Marginalized: Multiverse of Learning-Living in Grassroots Commons. *Educational Studies* 36(1): 58–73.

Provenzo, E. F. Jr. 2006. *Critical Issues in Education: An Anthology of Readings*. Thousand Oaks: Sage Publications.

Rahnema, M., & V. Bawtree (Eds). 1997. *The Post-Development Reader*. London: Zed Books.

Rassool, N. 1999. *Literacy for Sustainable Development in the Age of Information*. London, UK: Multilingual Matters Ltd.

Rees, M. 2003. *Our Final Century, Will the Human Race Survive the 21st Century?* London: William Heinemann.

Reiterman, T. 2005. National Environmental Policy Act Is "at a Crossroads." *Los Angeles Times*, July 7: 16.

Reitz, C. 2000. *Art, Alienation and the Humanities: A Critical Engagement with Herbert Marcuse*. Albany: SUNY Press.

Riley-Taylor, E. 2002. *Ecology, Spirituality, & Education: Curriculum for Relational Knowing*. New York: Peter Lang.

Roberts, P. 2003. Epistemology, Ethics and Education: Addressing Dilemmas of Difference in the Work of Paulo Freire. *Studies in Philosophy and Education* 22(2): 157–73.

———. 2000. *Education, Literacy, and Humanization: Exploring the Work of Paulo Freire*. Westport, CT:

Bergin & Garvey.

Robinson, W. I. 1996. Globalisation: Nine Theses on Our Epoch. *Race & Class* 38(2).

Rorty, A. O. 1998. Plato's Counsel on Education. In A. O. Rorty (Ed.) *Philosophers on Education.* New York: Routledge.

Rose, N. A. 1996. The Well-Being of Captive Marine Mammals: Concerns and Conflicts. In G. Burghardt, et. al. (Eds.), *The Well-Being of Animals in Zoo and Aquarium Sponsored Research.* Greenbelt, MD: Scientists Center for Animal Welfare.

Rosebraugh, C. 2004. *Burning Rage of a Dying Planet: Speaking for the Earth Liberation Front.* New York: Lantern Press.

Rothfels, N. 2002. *Savages or Beasts: The Birth of the Modern Zoo.* Baltimore: Johns Hopkins University Press.

Rovics, D. 2004. *David Rovics Songbook.* Randers, Denmark: Progressive Publishing: 148–49.

Royal Society. 1985. *The Public Understanding of Science.* London: Royal Society.

Russell, C., T. Sarick, & J. Kennelly. 2003. Queering Outdoor Education. *Pathways: The Ontario Journal of Outdoor Education* 15(1): 16–19.

Sale, K. 1985. *Dwellers in the Land.* San Francisco: Sierra Club Press.

Saltman, K. 2007. *Capitalizing on Disaster: Taking and Breaking Public Schools.* Boulder, CO: Paradigm Publishers.

Sandlin, J., & P. McLaren. (Eds.). 2009. *Critical Pedagogies of Consumption: Living and Learning in the Shopocalypse.* New York: Routledge.

Sandoval, C. 2000. *Methodology of the Oppressed.* Minneapolis, MN: University of Minnesota Press.

Sauvé, L. 2005. Currents in Environmental Education: Mapping a Complex and Evolving Pedagogical Field. *Canadian Journal of Environmental Education* 10(1): 11–37.

Sayers, D., & K. Brown. 1993. Freire, Freinet and "Distancing": Forerunners of Technology-Mediated Critical Pedagogy. *NABE News,* 17(3).

Scott, W., & S. Gough. 2004. *Key Issues in Sustainable Development and Learning: A Critical Review.* London: RoutledgeFalmer.

Seager, J. 2003. Rachel Carson Died of Breast Cancer: The Coming of Age of Feminist Environmentalism. *Signs: Journal of Women in Culture and Society* 28: 945–72.

Sears, P. B. 1964. Ecology: A Subversive Subject. *BioScience* 14: 11–13.

Selby, D. 2000. Humane Education: Widening the Circle of Compassion and Justice. In T. Goldstein & D. Selby (Eds.), *Weaving Connections: Educating for Peace, Social and Environmental Justice.* Toronto, ON: Sumach Press.

———. 1996. Relational Modes of Knowing: Learning Process Implications of a Humane and Environmental Ethic. In B. Jickling (Ed.), *A Colloquium on Environment, Ethics and Education*.Whitehorse, Yukon, Canada: Yukon College: 49–60.

———. 1995. *Earthkind: A Teacher's Handbook on Humane Education*. Stoke-on-Trent, UK: Trentham.

Selfe, C. L. 1999. *Technology and Literacy in the Twenty-First Century: The Importance of Paying Attention*. Carbondale: Southern Illinois University Press.

Semali, L., & J. Kincheloe. (Eds.). 1999. *What Is Indigenous Knowledge?: Voices from the Academy*. New York: Falmer Press.

Serres, M. 2001. *Le Tiers Instruit*. In M. Rahnema, with V. Bawtree (Eds.), *The Post-Development Reader*. London: Zed Books.

Shepard, P., & D. McKinley. 1969. *The Subversive Science: Essays Toward an Ecology of Man*. Boston: Houghton Mifflin.

Shiva, V. 2005. *Earth Democracy: Justice, Sustainability, and Peace*. Boston: South End Press.

———. 1997. Globalisation Killing Environment, Says Prominent Indian Green (From Frederick Noronha). May 30. Online at: http://www.hartford-hwp.com/archives/25a/061.html.

Singer, P., & J. Mason. 2006. *The Way We Eat: Why Our Food Choices Matter*. New York: Rodale.

Sleeter, C., & C. Grant. 2009. *Making Choices for Multicultural Education: Five Approaches to Race, Class and Gender* (6th ed.). Hoboken, NJ: Wiley.

Smith, L. T. 2002. *Decolonizing Methodologies*: Research and Indigenous Peoples. London and New York: Zed Books.

Snively, G., & J. Corsiglia. 2001. Discovering Indigenous Science: Implications for Science Education. *Science Education* 85(1).

Solnit, R. 2000. *Savage Dreams: A Journey into the Landscape Wars of the American West*. Berkeley: University of California Press.

Speth, J. G. 2009. *The Bridge at the Edge of the World: Capitalism, the Environment, and Crossing from Crisis to Sustainability*. New Haven, CT: Yale University Press.

Spretnak, C. 1999. *The Resurgence of the Real: Body, Nature, and Place in a Hypermodern World*. New York: Routledge.

Spring, J. 2004. *How Educational Ideologies are Shaping Global Society: Inter-Governmental Organizations, NGOs, and the Decline of the Nation-State*. Mahwah, NJ: Lawrence Erlbaum Associates.

Stanley, W. B., & N. W. Brickhouse. 2001. Teaching Sciences: The Multicultural Question Revisited. *Science Education* 85(1).

Stapp, W. 1969. The Concept of Environmental Education. *Journal of Environmental Education* 1(3): 31–36.

Steinfeld, H., P. Gerber, T. Wassenaar, V. Castel, M. Rosales, & C. de Haan. 2006. *Livestock's*

Long Shadow: Environmental Issues and Options. Food and Agriculture Organisation of the United Nations, Rome.

Sterling, S. 2001. *Sustainable Education: Re-Visioning Learning and Change*. Bristol, VT: Green Books.

Stone, A. R. 2001. Will the Real Body Please Stand Up? Boundary Stories about Virtual Cultures. In D. Trend (Ed.), *Reading Digital Culture*. Cambridge: MA. Blackwell Publishers.

Stone, M., & Z. Barlow. 2005. *Ecological Literacy: Educating Our Children for a Sustainable World*. San Francisco: Sierra Club Books.

Street, B. 1984. *Literacy in Theory and Practice*. Cambridge, UK: Cambridge University Press.

Stuchul, D. L., G. Esteva, & M. S. Prakash. 2005. From a Pedagogy of Liberation to Liberation from Pedagogy. In C. A. Bowers & F. Apffel-Marglin (Eds.), *Rethinking Freire: Globalization and the Environmental Crisis*. Mahwah, NJ: Lawrence Erlbaum Associates.

Suoranta, J., & T. Vaden. Forthcoming. *Wikiworld*. London: Pluto Press.

Suppes, P. 1968. Computer Technology and the Future of Education. *Phi Delta Kappan*. April: 420–23.

Swimme, B., & T. Berry. 1994. *The Universe Story: From the Primordial Flaring Forth to the Ecozoic Era—A Celebration of the Unfolding of the Cosmos*. San Francisco: Harper.

Tarnas, R. 1991. *The Passion of the Western Mind: Understanding the Ideas That Have Shaped Our World View*. New York: Ballantine.

Teilhard de Chardin, P. 1965. *The Phenomenon of Man*. New York: Harper and Row.

Todd, R. D. 1991. The Natures and Challenges of Technological Literacy. In M. J. Dyrenfurth & M. R. Kozak (Eds.), *Technological Literacy*. Peoria, IL: Glencoe, McGraw-Hill: 10–27.

Torres, C. A. 2004. Els Mons Distorsionats de Paulo Freire i Ivan Illich. *diàleg: Paulo Freire i Ivan Illich*. Xàtiva, Spain: Centre de Recursos i Educació Continua.

Trend, D. 2001. *Welcome to Cyberschool: Education at the Crossroads in the Information Age*. Lanham, MD: Rowman & Littlefield.

Tyack, D. 2000. Reflections on Histories of U.S. Education. *Educational Researcher* 29(8).

United Nations Conference on Environment and Development. 1992. Promoting Education, Public Awareness and Training. In *Agenda 21*. Geneva: UN: 221–27.

United Nations Environment Programme. 2002. *Global Environmental Outlook 3: Past, Present, and Future Perspectives*. Geneva, UN: 13–15.

UNESCO. 1999. *Science and Technology Education: Philosophy of Project 2000+*. The Association for Science Education. Paris, France: UNESCO.

———. 1994. *The Project 2000+ Declaration: The Way Forward*. Paris, France: UNESCO.

Bibliography

United Nations. 1992. *Report of the United Nations Conference on Environment and Development.* Rio de Janeiro, Brazil: UNCED.

U.S. Congress. 2001. *No Child Left Behind Act of 2001.* Public Law 107–110. Washington, DC.

U.S. Department of Education. 2004. *Toward a New Golden Age in American Education: How the Internet, the Law, and Today's Students Are Revolutionizing Expectations.* Washington, DC: National Education Technology Plan.

———. 1996. *Getting America's Students Ready for the 21st Century—Meeting the Technology Literacy Challenge, A Report to the Nation on Technology and Education.* Washington, DC: National Education Technology Plan.

Verrengia, J. B. 2003. Scientists Raise Alarm Over Sea-Mammal Deaths. *The Associated Press* (June 16).

Waetjen, W. 1993. Technological Literacy Reconsidered. *Journal of Technology Education* 4(2): 5–11.

Wagner, D., & R. Kozma. 2003. New Technologies for Literacy and Adult Education: A Global Perspective. Paper for NCAL/OECD International Roundtable. Philadelphia, PA. Online at: http://www.literacy.org/ICTconf/PhilaRT_wagner_kozma_final.pdf.

Wajcman, J. 2004. *Technofeminism.* Malden, MA: Polity Press.

Waldram, J. B. 1993. Aboriginal Spirituality in Corrections: A Canadian Case Study in Religion and Therapy. *American Indian Quarterly* 18(2): 197–215.

Watson, P. 1995. The Cult of Animal Celebrity. *Animal People.* June.

Wayne, K. W., & D. A. Gruenewald. (Eds.). 2004. Special Issue: Ecojustice and Education. *Educational Studies* 36(1).

Weber, M. 1958. *The Protestant Ethic and the Spirit of Capitalism.* Trans. Talcott Parsons. New York: Charles Scribner's Sons.

Weil, D. K. 1998. *Toward a Critical Multicultural Literacy.* New York: Peter Lang.

Weil, Z. 2004. *The Power and Promise of Humane Education.* Gabriola Island, BC: New Society Publishers.

Weiss, R. 2003. Key Ocean Fish Species Ravaged, Study Finds. *Washington Post*, May 15.

Wells, H. G. 1938. *World Brain.* New York: Doubleday.

Wenden, A. (Ed.). 2004. *Educating for a Culture of Social and Ecological Peace.* Albany: SUNY Press.

White, L. 1996. The Historical Roots of our Ecological Crisis. In R. S. Gottlieb (Ed.), *This Sacred Earth.* New York: Routledge.

Whitehead, A. N. 1970. *Science and the Modern World.* New York: The Free Press.

Wilson, K. 2003. Therapeutic Landscapes and First Nations Peoples: An Expoloration of Culture, Health, and Place. *Health & Place* 9: 83–93.

Wolhuter, C. C. 2001. History of Environmental Education: Lacuna on the Research Agenda of History of Education. Birmingham, UK. Online at: http://216.239.39.100/search?q =cache:c-OhlGLTM0cC:www.cetadl.bham.ac.uk/ische/abstract/WolhunterA.doc.

World Summit on the Information Society. 2003. *Plan of action*. Document WSIS-03/GENEVA/ DOC/5-E (December 12). Online at: http://www.itu.int/wsis/.

Worm, B., et al. 2006. Impacts of Biodiversity Loss on Ocean Ecosystem Services. *Science* 314(5800): 787–90.

Wyer, M., & M. Barbercheck, D. Giesman, H. Ö. Öztürk, & M. Wayne. 2001. *Women, Science, and Technology*. New York: Routledge.

Yang, S. K., & R. Hung. 2004. Towards Construction of an Ecopedagogy Based on the Philosophy of Ecocentrism. *Journal of Taiwan Normal University* 49(2).

ZDNet. 2001. More News: Why Ballmer's "Monkey Boy" Dance Was a Tour de Force. August 24. Online at: http://review.zdnet.com/4520-603316-4206342.html.

Zerzan, J. 2002. *Running on Emptiness: The Pathology of Civilization*. Los Angeles: Feral House.

Zizek, S. 1999. Attempts to Escape the Logic of Capitalism. *London Review of Books*, 21(21).

Index

A Nation at Risk, 67–68
Abbey, Edward, 146
academic freedom, 11
Adler, Mortimer, 44–45, 59
Agamben, Giorgio
    concept of "anthropological machine", 45
Alaska Native Rural Systemic Initiative, 105
alliance politics, 14, 27, 57, 109, 115, 146
alter-globalization, 16, 86, 115, 126, 140
Anderson, Perry, 15
Andrzejewski, Julie, 11, 31, 155
Animal Liberation Front, 126, 127
animal rights, 8, 19, 28, 109, 132, 134
anthropocentrism, 9, 19, 152
    critiques of, 13, 22–23, 58, 139
anti-globalization. *See* alter-globalization
Apple, Michael, 25
Aristotle, 41, 49–51, 57
Aronowitz, Stanley, 25, 68, 82
Bahro, Rudolf, 17
Bari, Judi, 145–148
Berry, Thomas, 36, 46, 59
Best, Steven, 3, 7, 11, 35, 67, 84, 108, 126–127, 132, 143
    concept of " interspecies alliance politics", 126
biophilia, 18, 22, 32
bios, 136–137
    biocide, 136
Boff, Leonardo, 19

Bookchin, Murray, 1, 4, 40, 101, 143, 146
    concept of "ecology of freedom", 108, 139, 143
Bowers, C. A., 20, 47, 82
Burbules, Nick, 75
Bush administration, 10, 29, 54, 69, 125, 131, 135
Cajete, Gregory, 32, 105, 118, 122
Calderon, Dolores, 112
Camara, Dom Helder, 19, 101
capitalism, 30, 32, 46, 52, 55, 70, 77, 81, 88, 99, 113, 118, 131, 136, 140, 145
    ecological costs of, 2, 3, 30
    neoliberalism, 3, 16, 18, 21, 54, 63, 88, 105, 121, 125, 135, 140, 148
    technocapitalism, 3, 9, 57, 62, 67–68, 84, 87, 90, 92, 109, 114, 152
    Third Way, 15, 30
Capra, Frijtof, 4, 11
Carson, Rachel, 1, 129, 147
Castells, Manuel, 62, 82, 84, 87
Chávez, César, 145
Cicero, 44, 59, 163
citizenship. *See* cosmopolitan
climate change. *See* global warming
Clinton, Bill, 30, 63, 68
    Clinton Global Initiative, 15–6
cognitive praxis, 25, 26, 27, 33, 112, 148
    as epistemic standpoint, 26, *See* standpoint
Collins, Patricia Hill
    concept of "matrix of domination", 9

commons, 11, 55, 59, 80, 99, 103, 105, 113, 115–117, 120, 123
communion, 36, 47, 84
cosmology
   Aristotelian chain of being, 50
   definition of, 35–36
   earthling paradigm, 56
   Ecozoic era, 36, 46
   Newtonian-Cartesian paradigm, 52, 108
   Platonic turn, 49
   Western worldview, 35
cosmopolitan, 43, 55, 113
   citizenship, 45, 57, 99, *See* planetarity
crisis, idea of, 4
Darder, Antonia, 31, 85, *See* Preface
Decade of Education for Sustainable Development, 13, 15, 17, 103
Deloria, Vine, 59, 105, 107, 122, 143
democracy. *See* ecological democracy
Descartes, René, 52
Dewey, John, 53, 61, 64, 83
direct action, 32, 109, 130, 131, 140
Earth Charter, 12, 13, 18, 21
Earth Day, 1, 128
Earth Liberation Front, 126, 127, 128, 130, 137, 138, 140, 141
Earth Summit, 12, 13, 18, 31, 70
ecocrisis. *See* ecological crisis
ecofeminism, 146
ecoliteracy, 19, 23, 26, 29, 59
   as social and cultural literacy, 11
   cosmological dimension of, 26
   depoliticization of, 140
   educational frameworks for, 11–12
   organizational dimension of, 26
   necessity of, 5–6, 18, 27, 46, 62, 82, 98, 118–119
   technological dimension of, 26
ecological crisis, 4–5, 24, 36, 78, 127
   causes of, 9, 22, 30
ecological democracy, 57, 59, 74, 115, 118, 120, 149
Ecological Society of America's SEEDS program, 105
eco-modernization, 28
ecopedagogy movement
   as critical theory of technology, 63
   as education for sustainability, 29, 106
   as form of critical pedagogy, 18
   as international movement, 18
   as oppositional movement, 19, 23, 63, 72
   as standpoint methodology, 112
   challenge to environmental studies, 104
   definition of, 18, 22, 26–27, 30
   dialectical and collaborative, 25
   eco-critique of critical pedagogy, 20
   for alliance politics, 57
   for the global north, 20–22, 27
   Marcusian form of, 127, 136
   post-anthropocentric nature of, 22
   relationship to traditional ecological knowledge and WMS, 107, 112, 118, 123
   situational nature of, 21
ecosocialism, 57
ecovillage, 115
education
   as *educare* and *educere*, 101
   as *paideia*, 39, *See paideia*; *humanitas*
   commons-based, 11
   critical pedagogy, 8, 18–23, 25, 30–31, 72, 82, 87, 99, 109, 138
   eco-justice, 11, 47, 58, 126
   ecological, 11
   ecopedagogy. *See* ecopedagogy movement
   environmental, 5–11, 13, 16–18, 27–29, 140
   holistic, 11
   humane, 11
   media culture as, 73, 82
   multicultural. *See* multicultural education
   outdoor education, 7, 28
   peace, 11

place-based, 11
popular education, 5, 19
schools as sites of struggle, 21
science. *See* science
traditional ecological knowledge. *See* traditional ecological knowledge (TEK)
transformative, 11
education for sustainable development, 12–14, 16–17, 19, 28–29
  as non-participatory and instrumentalist, 14
  as "disaster capitalism", 17
  interstitial tactic, 14
Eisler, Riane
  concept of "Dominator Hierarchies", 48
Empire, 113–117
environmental crisis. *See* ecological crisis
environmental justice, 7, 14, 31
environmental literacy, 11, 18
  definition of, 29
  depoliticization of, 10
  example of. *See* Zoo School
  lack of, 6
  limitations of, 5, 7–10
  standards-based versions, 8–9
environmental movement, 1, 5–6, 19, 26, 97, 116, 126, 128, 131, 134, 141
  knowledge interests of, 26, *See* cognitive praxis
environmental science, 103, 105
  as form of WMS, 105
environmental studies, 103–106, 108, 121
  as environmental science, 6, 104
  marginalization of indigenous perspectives, 106
ethnoscience, 91, 101, 111
Exxon Valdez, 131, 142
factory farms, 3
Feenberg, Andrew, 76
Food Not Bombs, 116
Frankfurt School, 22, 47, 95, 141, 143
Freire, Paulo, 18–19, 37, 71, 79, 81–82, 87, 89, 93, 101, 138, 145
  concept of " conscientization", 25, 91–92, 101
  concept of "cultural invasion", 90
  critical emancipatory literacy, 66
  cultural circle and technology, 89
  education as politics, 22
  humanization as speciesist idea, 21
  liberation theology connections, 19
  on ecopedagogy, 20–21
  problematization of technology, 87
  promethean philosophy, 23, 87
Fromm, Erich, 21
Gadotti, Moacir, 18, 91
Galeano, Eduardo, 125
Get Oil Out!, 129
Giroux, Henry, 22, 25, 31
global warming, 2, 10, 16, 28, 30, 132, 144
globalization, 2–3, 6, 15, 18, 21, 36, 39, 41, 46, 52, 54, 59, 66, 88, 96, 98, 113, 119, 136
Goldman, Emma, 81
González-Gaudiano, Edgar, 13–15, 19
Gore, Al, 2, 28, 68, 132
Gramsci, Antonio, 83, 142
  politics as education, 22
Grande, Sandy, 31, 107, 118–120
green anarchism, 57
  anarcho-primitivism, 127
Gruenewald, David, 11, 13, 28, 31, 80
Gur-Ze'ev, Ilan, 20, 127, 138
Hammer, Rhonda, 72, 74, 76, 79
Haraway, Donna, 59, 112, 164
Harding, Sandra, 6, 78–79, 101, 104, 105–106, 112, 121–123
Hardt, Michael, 57, 113–115, 118–119
Harney, Corbin, 108–109
Hellenistic dispersal, 41–43
Hill, Julia Butterfly, 59, 146
Homo educandus. *See* Homo educans
  Illichian idea of, 93
Homo educans, 37, 47
Horton, Myles, 21

humanitas, 36, 44–45, 47–48, 51–52, 100, 128, *See* Adler, Mortimer
  connection to humanism, 48
  conservative and reactionary aspects of, 44
  Marcusian conception of, 54, 138
Hurricane Katrina, 131
ideology of dehumanization, 100
Illich, Ivan, 14, 21, 24, 30, 64, 79, 81, 93, 98, 145, 147
  as epimethean philosopher, 24, 93
  as post-development thinker, 24
  concept of "Techno-Moloch", 96
  convivial tools, 64, 96
  definition of counterproductive tools, 95
  definition of education, 94
  concept of "global classroom", 14, 30
  Internet as an Illichan "learning web", 98
  relationship to Freire, 25
imperialism, 3, 18, 21, 43, 104, 119, 130
inclusive democracy, 43
indigenous peoples, 31, 105, 106, 112, 119
  characterized as " ecological Indians", 106
information society, 3, 84, 97
Jensen, Robert, 58
Jickling, Bob, 14–15
Kellner, Douglas, 3, 22–23, 32, 35–36, 54–55, 58, 62, 67, 72–76, 78–79, 82–84, 86, 100, 108, 127, 132–135, 138–143, *See* Afterword
  concept of "diagnostic critique", 83
Kincheloe, Joe, 30–31, 123
LaDuke, Winona, 105, 109, 119
Landless Rural Workers' Movement, 19
Lennox School District, CA, 85
literacy
  definition of, 65
Lorde, Audrey, 94
Lyons, Oren, 120
Macedo, Donaldo, 66, 73

Marcos, Subcomandante Insurgente, 148
Marcuse, Herbert, 21–22, 54, 95, 100, 108, 127, 132, 137, 145
  concept of "Great Refusal", 23–24, 134
  concept of "conceptual mythologies", 23
  concept of "objective ambiguity", 83
  concept of " one-dimensional", 95, 136
  concept of "ecocide", 135
  concept of "repressive tolerance", 132
  domination of nature, 22, 48, 139
  revolutionary violence, 133
  theory of education, 23, 54, 138
Martusewicz, Rebecca, 11
Marx, Karl, 23, 46–47, 79, 88, 94
mass extinction, 2–3, 10, 30, 36, 47, 54, 57, 125, 135, 139
McKenzie, Marcia, 60
McLaren, Peter, 11, 14, 31, 79, 87, 100, 138, 145
McLuhan, Marshall, 61, 65, 101
media spectacle, 28, 61, 74, 78, 84, 87, 129, 134, 142
  megaspectacle, 85
Millennium Ecosystem Assessment, 2
Morales, Evo, 31
Morin, Edgar, 31
multicultural education, 72–73, 98, 106, 112, 118
  culturally relevant forms of, 18, 105
multitude, 28, 113, 114, 118, 119, 120
  aesthetic dimension of, 117
  as temporary autonomous zone, 117
  Peace Camp as example of, 115
Negri, Antonio, 57, 113–115, 118–119
neoliberalism. *See* capitalism
Nevada Test Site, 107–108
new science, 59, 108, 113–115, 118, 120, 143
  Marcusian, 55, 108, 139
  of the multitude, 114, 119
Nietzsche, Friedrich, 1, 57
Nkrumah, Kwame, 101

Index

No Child Left Behind Act, 9, 63, 69
North American Association for Environmental Education, 5, 8, 29
O'Sullivan, Edmund, 46
Obama, Barack, 29–30, 119, 125, 131, 140
Orr, David, 11, 76
paideia, 36, 38–45, 49, 51–53, 56–59, 114, *See* Adler, Mortimer
    as civic literacy, 37
    as democratic principle, 40, 42
    Athenian history of, 38–39, 41, 43, 48
    contradictions of, 41
    ecological, 37, 47–48, 54, 57, *See* ecological democracy
    global phase, 54
    role in cultivating civilization, 40, 48
    role in Hellenistic dispersal, 42
Papert, Seymour, 92
patriarchy, 21, 23–24, 32, 49, 104, 147
Pena, Devon, 121
Petrina, Stephen, 63
planetary, 17–18, 47, 55, 61, 77, 83, *See* cosmopolitan
    planetary citizenship, 46, 48, 54, 58, 87, 99, 107
Plato, 41, 43, 49–51, 56, 143
    Allegory of the Cave, 56
political imagination, 23, 30, 118, 139, 142
Postman, Neil, 62, 73
Prakash, Madhu Suri, 11, 24, 59, 82
Project 2000+, 63, 70
    scientific and technological literacy, 70, 71, 79
racism, 14, 21, 104, 111
    white privilege, 59
repressive tolerance, 11, 133, 142, *See* Marcuse, Herbert
Rovics, David, 130
Saxon, Levana, 31
science, *See* new science
    as WMS. *See* western modern science (WMS)

definition of, 111
    democratic forms of, 70, 112, 110, 118
    education, 9
Seager, Joni, 147
Shiva, Vandana, 59, 103
Shundahai Network, 107–108
Snively, Gloria, 105, 111, 121
social justice, 19, 56, 66, 71, 93, 96–97, 109, 130
social movements, 17, 22, 27, 62, 70, 82, 126–127, 134, *See* ecopedagogy; environmental movement
    student movement, 32, 131
social reconstruction, 3, 7, 11, 17, 21, 27, 32, 36, 39, 63, 66, 72, 76–78, 83, 92, 96, 98–100, 108, 112, 114, 128, 134, 141, 148
    global north; south, 19–20, 31, 87, 148
    Green and Brown agendas, 31
sovereignty, 119
    relationship to democracy, 119
standpoint, 78, 107, 112, *See* cognitive praxis
Stapp, William, 7, 10
Steinberg, Shirley, 73, 122
sustainable development, 12–14, 16–17, 29–30, 46, 63, 70, 121
    definition of, 15
Swimme, Brian, 36, 58
technoliteracy, 9, 61–63, 65–67, 69–73, 76–79, 85
    as social and community-building practice, 70
    as technical and vocational competencies, 69
    computer literacy, 62, 68–70, 74, 82, 97
    critical media literacy, 72–74, 76, 82, 88
    critical multimedia literacy, 76
    concept of "multiliteracies", 67
    multiple literacies, 63, 67, 71, 75–76, 98–99
technology

appropriate, 5, 24, 63, 65, 67, 71, 77, 96, 101
ARPANET, 67
  definition of, 63
  information-communication technologies (ICTs), 62–65, 67, 70–71, 77, 79, 86, 91, 107
  sustainable design, 76
terror, 9, 16, 86, 96, 112, 126, 132, 147
Third Way economics. *See* capitalism, Third Way
Third World, 79, 90–91, 101, 104, *See* global north; south
traditional ecological knowledge (TEK), 32, 57, 60, 106
  academic marginalization of, 106
  as sustainable praxis, 106
  contemporary versions of, 107
  definition of, 105
  grounded in cosmological conditions, 105
  sweat lodge as example of, 107, 109–111, 117, 122

United Nations Environment Programme, 1, 3, 12, 19
veganism, 8, 116, 144
war, 38, 86, 113, 125–126, 130–131, 134–135, 140
Weber, Max, 55, 95
Western civilization, 22, 37, 38, 43, 47, 48, 56
western modern science (WMS)
  institutional science as WMS, 104, 110
  continuity with traditional ecological knowledge, 105
  definition of, 112
  industrial origins of, 111
Whitehead, Alfred North, 35, 56
Wildcat, Daniel, 59, 105, 107, 122, 143
Wobbly, 146, 148
World Summit for Sustainable Development, 13, 29
Zizek, Slovoj, 30
zoë, 136–137, 143
zoöcide, 136
Zoo School, 7–9